THE HUMAN GENOME

THE MEDICAL PERSPECTIVES SERIES

Editors:

Andrew P. Read *Department of Medical Genetics, University of Manchester, St Mary's Hospital, Hathersage Road, Manchester M13 0JH, U.K.*

Terence Brown *Department of Biochemistry and Applied Molecular Biology, UMIST, Manchester M60 1QD, U.K.*

Oncogenes and Tumor Suppressor Genes

Cytokines

The Human Genome

Autoimmunity (due early 1992)

Genetic Engineering (due early 1992)

THE HUMAN GENOME

T. Strachan
University Department of Medical Genetics, St Mary's Hospital, Hathersage Road, Manchester M13 0JH, U.K.

*β*IOS
SCIENTIFIC
PUBLISHERS

© BIOS Scientific Publishers Limited, 1992

First published in the United Kingdom 1992 by
BIOS Scientific Publishers Limited,
St Thomas House, Becket Street, Oxford OX1 1SJ.

A CIP catalogue record for this book is available from the British Library.

ISBN 1 872 748 80 5

To my family and the memory of Hugh M. Strachan, 1947–1979

Typeset by Enset Photosetting Limited, Bath, U.K.
Printed by Information Press Ltd, Oxford, U.K.

PREFACE

The last few years have witnessed extraordinary advances in our understanding of the human genome. The genes which underlie many important inherited disorders such as cystic fibrosis and Duchenne muscular dystrophy have recently been isolated and studied, as have many genes which cause human cancers. These developments have led to improved diagnosis of genetic disease and to a greatly increased understanding of the molecular basis of single gene disorders.

Currently, much effort is being devoted to identifying genes implicated in the pathogenesis of common multifactorial disorders such as cancer, heart disease, mental illness, etc. In addition, gene technology has begun to be applied to devising new treatments for human disease. Against this background, the Human Genome project, one of the most ambitious scientific endeavors ever undertaken, has recently been initiated with the ultimate aim of isolating and characterizing each of the 50 000 to 100 000 genes in the human genome. The aim of this book is to provide non-specialist and specialist alike with a concise description of our current knowledge of the human genome and the ways in which it is influencing medical research and practice.

My thanks go to the series editors, Drs Andrew Read and Terry Brown for their helpful comments on the manuscript, and to my colleagues Paul Sinnott, Paul Sinclair, Carolyn Watson and Andrew Wallace for providing photos reproduced in Figures 3.8, 4.9, 4.10 and 6.8, respectively.

T. Strachan

CONTENTS

ABBREVIATIONS

ADA	adenosine deaminase
APC	adenomatous polyposis coli
ARMS	amplification refractory mutation system
ARS	autonomously replicating sequence
ASO	allele-specific oligonucleotide
bp	base pair
CF	cystic fibrosis
CFTR	cystic fibrosis transmembrane regulator
cM	centimorgan
DGGE	denaturing gradient gel electrophoresis
DMD	Duchenne muscular dystrophy
FAP	familial adenomatous polyposis
FISH	fluorescence *in-situ* hybridization
HLA	human leukocyte antigen complex
HUGO	Human Genome Organization
Ig	immunoglobulin
kb	kilobase
LCR	locus control region
LDL	low density lipoprotein
LINE	long interspersed nuclear element
Mb	megabase
mRNA	messenger RNA
MHC	major histocompatibility complex
mt	mitochondrial
NF1	neurofibromatosis type I
PCR	polymerase chain reaction
PFGE	pulsed field gel electrophoresis
PIC	polymorphism information content
rDNA/RNA	ribosomal DNA/RNA
RFLP	restriction fragment length polymorphism
RSP	restriction site polymorphism
SINE	short interspersed nuclear element
SnRNA	small nuclear RNA
SSCP	single-strand conformation polymorphism
STS	sequence-tagged site
TIL	tumor infiltrating lymphocytes
TNF	tumor necrosis factor
tRNA	transfer RNA
VNTR	variable number of tandem repeats
YAC	yeast artificial chromosome

1
ORGANIZATION AND EXPRESSION OF THE HUMAN GENOME

1.1 Structure of genomic DNA

The human genome consists of DNA molecules in the form of a double helix in which the two strands of the DNA duplex are held together by weak hydrogen bonds. Each strand has a linear backbone of residues of deoxyribose (a 5-carbon sugar) which are linked by covalent phosphodiester bonds. Covalently attached to carbon atom number 1' of each sugar residue is a nitrogenous base, either a pyrimidine (cytosine or thymine), or a purine (adenine or guanine; see *Figure 1.1*). A sugar with an attached base and phosphate group therefore constitutes the basic repeat unit of a DNA strand, a nucleotide. As the phosphodiester bonds link carbon atoms number 3' and number 5' of successive sugar residues, one end of each DNA strand, the so-called 5' end, will have a terminal sugar residue in which carbon atom number 5' is not linked to a neighboring sugar residue. The other end is defined as the 3' end because of a similar absence of phosphodiester bonding at carbon atom number 3' of the terminal sugar residue. The two strands of a DNA duplex always associate (anneal) in such a way that the 5'→3' direction of one DNA strand is the opposite to that of its partner.

Genetic information is encoded by the sequence of bases in the DNA strands. Hydrogen bonding occurs between laterally opposed bases, base pairs, of the two strands of a DNA duplex according to Watson–Crick rules: adenine (A) specifically binds to thymidine (T) and cytosine (C) specifically binds to guanine (G). Consequently, the two strands of a DNA duplex are said to be complementary (or to exhibit base complementarity) and the sequence of bases of one DNA strand can readily be determined if the DNA sequence of its complementary strand is already known. It is usual, therefore, to describe a DNA sequence by writing the sequence of bases of one strand only, and in the 5'→3' direction, which is the direction of synthesis of new DNA molecules during DNA replication and also of transcription when RNA molecules are synthesized using DNA as a template. However, when describing the sequence of a DNA region encompassing two neighboring bases (really a dinucleotide) on one DNA strand, it is usual to insert a 'p' to denote a connecting phosphodiester bond. For example, CpG means that a cytosine is covalently linked to a neighboring guanine on the same DNA strand, while a CG base pair means a cytosine on one DNA strand is hydrogen bonded to a guanine on the complementary strand.

In the process of DNA replication, the two DNA strands unwind and each strand directs the synthesis of a complementary DNA strand to generate two daughter DNA

1

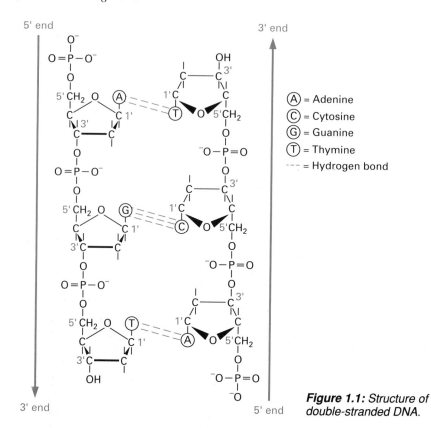

5' end

3' end

(A) = Adenine
(C) = Cytosine
(G) = Guanine
(T) = Thymine
- - - = Hydrogen bond

3' end

5' end

Figure 1.1: *Structure of double-stranded DNA.*

duplexes which are identical to the parent molecule. However, during gene expression, individual genes are transcribed from only one of the two available DNA strands. The strand from which the RNA is transcribed, and which is complementary in base sequence to the RNA molecule, is the so-called template strand (or anti-sense strand). The transcribed single-stranded RNA molecule is therefore a faithful copy of the other DNA strand of the DNA duplex, the sense strand, except that the sugar in the RNA molecule is ribose, and uracil (U) bases replace the original thymines. In the case of documented gene sequences it is customary to show only the DNA sequence of the sense strand. Orientation of sequences relative to a gene sequence is commonly dictated by the sense strand and by the direction of transcription. For example, the 5' end of a gene refers to the DNA sequence at the 5' end of the sense strand, and sequences upstream or downstream of a gene refer to sequences which flank the gene at the 5' or 3' ends of the sense strand respectively.

1.2 The nuclear and mitochondrial genomes

The genetic information in human cells is organized in the form of two genomes, a complex nuclear genome and a simple mitochondrial genome. The difference in complexity of the two genomes reflects the predominance of the nuclear genome in providing the great bulk of essential genetic information, most of which is ultimately decoded to specify polypeptide synthesis on cytoplasmic ribosomes. Although mitochondria possess their own ribosomes, the mitochondrial genome specifies only a very small portion of the specific mitochondrial functions; the bulk of the mitochondrial polypeptides are encoded by nuclear genes and are imported from the cytoplasm.

1.2.1 The nuclear genome

The nuclear DNA content of individual human cells is determined by the number of nuclei and the number of chromosomes in that cell. The specialized germ cells, eggs and sperm cells, are haploid cells in which there is a single copy of the nuclear genome with the DNA being distributed between 23 chromosomes, comprising 22 autosomes and a single sex chromosome, X or Y. Fusion of a normal egg cell and sperm cell at conception generates a diploid zygote with two genome copies (2C) and 46 chromosomes, consisting of 23 pairs of homologous chromosomes, that is, an homologous pair of each of the 22 autosomes and two sex chromosomes which may be completely homologous (XX), or partially homologous (XY).

Table 1.1: *DNA content of human chromosomes*[a]

Chromosome	Percentage of total length	Amount of DNA (Mb)	Chromosome	Percentage of total length	Amount of DNA (Mb)
1	8.3	250	13	3.6	110
2	7.9	240	14	3.5	105
3	6.4	190	15	3.3	100
4	6.1	180	16	2.8	85
5	5.8	175	17	2.7	80
6	5.5	165	18	2.5	75
7	5.1	155	19	2.3	70
8	4.5	135	20	2.1	65
9	4.4	130	21	1.8	55
10	4.4	130	22	1.9	60
11	4.4	130	X	4.7	140
12	4.1	120	Y	2.0	60

[a]The DNA content is given for chromosomes prior to entering the S (DNA replication) phase of cell division (see *Figure 1.3*); data abstracted from reference 1.

Subsequent mitotic DNA duplication and cell division events during development and growth result in the great majority of somatic cells containing a single diploid nucleus. Exceptions include examples of naturally polyploid cells in which there are additional rounds of chromosome duplication prior to cell division (e.g. certain liver cells normally each have 92 chromosomes and are tetraploid, i.e. have four copies of the haploid genome) and cells which are multinucleated (e.g. fully differentiated muscle cells) or which lack a nucleus (e.g. mature red blood cells). Additionally, although the number of chromosomes in a normal diploid somatic cell remains 46 until the anaphase stage of mitosis, the cell becomes effectively tetraploid following DNA duplication during the earlier S phase of the cell cycle as described below.

By and large, somatic cells carry the same genetic information as the zygote from which they arise. Except for sequences in the non-homologous regions of the X and Y chromosomes in males (see Section 2.2), diploid cells contain two copies of each individual nuclear gene (or DNA sequence) normally present in haploid cells. A pair of such homologous DNA sequences, which are located at identical positions on homologous chromosomes (i.e. at the same locus) are referred to as alleles. Most nuclear DNA sequences show Mendelian inheritance, whereby in diploid cells one allele is inherited from each parent. An individual is said to be homozygous or heterozygous at a specific locus if the two alleles at that locus are, respectively, identical or different in sequence. As Y chromosomes are transmitted exclusively by males, Y chromosome sequences are present as single copies in male diploid cells, as are X chromosome sequences.

The nuclear genome of a human haploid cell contains about 3×10^9 bp of DNA and an average size chromosome has approximately 1.3×10^8 bp (or 130 megabases) of DNA but can vary between approximately 50 Mb and 250 Mb (see *Table 1.1*). The DNA content of each chromosome is thought to consist of a single linear double-stranded DNA molecule which, if fully uncoiled, would be between 1.7 and 8.5 cm long. In the cell the structure of each chromosome is highly ordered [2] and compaction of the chromosomal DNA is achieved by complexing with various DNA-binding proteins. The most fundamental unit of packaging is the nucleosome which consists of a central core complex of eight basic histone proteins (two each of histones H2A, H2B, H3 and H4) around which a stretch of 146 bp of double-stranded DNA is coiled in 1.75 turns (*Figure 1.2*). Adjacent nucleosomes are connected by a short length of spacer DNA. The elementary fiber of linked nucleosomes is in turn coiled into a chromatin fiber of 30 nm diameter which can be resolved by electron microscopy.

At the metaphase stage of cell division the chromosomes become even more condensed and can be resolved by optical microscopy as structures which are over 1 μm wide and range in length from 2 μm (chromosome 21) to 10 μm (chromosome 1). At this stage the DNA in the chromosome arms, but not that in the centromere, has already duplicated in preparation for cell division. The metaphase chromosome consists of two laterally opposed chromatids which remain bound to each other at the centromere. Each chromatid consists of loops of chromatin fiber, containing approximately 30–90 kb of DNA per loop, which are attached to a central scaffold of non-histone acidic protein and the resulting complex is further compacted by coiling. As each chromatid contains double-stranded DNA (except for the centromere DNA), the genomic DNA in a human somatic cell from late S phase right up to the anaphase stage of mitosis is effectively tetraploid (4C), although the chromosome number is still 46 (*Figure 1.3*).

A variety of treatments cause chromosomes in dividing cells to appear as a series of alternating light and dark staining bands. In G-banding, for example, the chromosomes are subjected to controlled digestion with trypsin before staining with the DNA-binding chemical Giemsa which reveals alternating positively (dark G-bands) and negatively staining regions (pale G-bands). Bands are classified according to their relative location on the short arm (p) or long arm (q) of specific chromosomes. For example, 17p12 means sub-band 2 of band 1 on the short arm of chromosome 17. Further sub-division is also possible: 17q21 can be divided into 17q21.1, 17q21.2 and 17q21.3 (*Figure 1.2*). As Giemsa shows preferential binding to A + T rich DNA

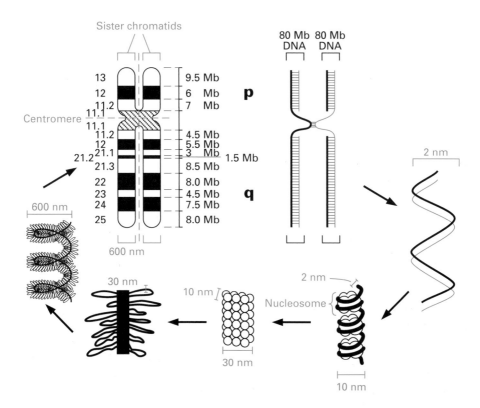

Figure 1.2: *From DNA duplex to metaphase chromosome (human chromosome 17, Giemsa-stained, 550 band preparation).*

sequences, the dark G-bands have been considered to be rich in A + T bases, while the pale G-bands are rich in G + C. Although, therefore, the average base composition of human genomic DNA is about 40% G + C, the alternating pale and light bands are thought to reflect the compartmentalization of the human genome into isochores, defined chromosomal regions in which the base composition of the DNA is comparatively homogeneous but which is variable between isochores [3]. The dark G-bands are thought to be deficient in genes (see below); during the cell cycle these regions condense early but replicate late. In contrast, pale G-bands represent regions which condense late, but replicate early; they are rich in G + C and in genes. Additionally, the two types of chromosomal regions differ in their predominant association with particular classes of interspersed repetitive DNA (see Section 1.8.4). At the resolution of approximately 550 bands per set of mitotic metaphase chromosomes (karyogram– see *Figure 1.4*), an average size band corresponds to approximately 6 Mb of DNA.

In the haploid nuclear genome the total number of genes is thought to be approximately 50 000–100 000. On this basis all nucleated cells have, on average, one gene per 30–60 kb, and about 2000–4000 per average chromosome. In a 400-band mitotic metaphase karyogram, one might anticipate about 100–200 genes on average per band. However, as noted above, average gene density is dependent on the base composition of the chromosomal region containing the gene and pale G-bands are relatively enriched in genes at the expense of dark G-bands.

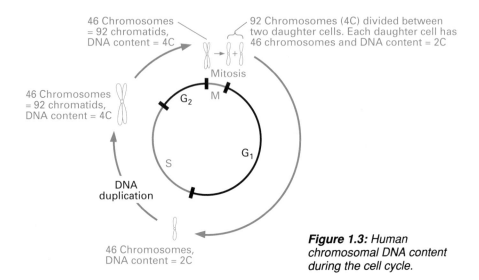

46 Chromosomes
= 92 chromatids,
DNA content = 4C

92 Chromosomes (4C) divided between
two daughter cells. Each daughter cell has
46 chromosomes and DNA content = 2C

Mitosis

46 Chromosomes
= 92 chromatids,
DNA content = 4C

G_2

M

S

G_1

DNA
duplication

46 Chromosomes,
DNA content = 2C

Figure 1.3: *Human chromosomal DNA content during the cell cycle.*

1.2.2 The mitochondrial genome

The mitochondrial genome is defined by a single type of circular double-stranded DNA molecule, 16 569 bp long, which has been completely sequenced [4]. During zygote formation a sperm cell contributes its nuclear genome, but not its mitochondrial genome, to the egg cell. Consequently, the mitochondrial genome of the zygote is determined exclusively by that originally found in the unfertilized egg. The mitochondrial genome is therefore maternally inherited; males and females both inherit their mitochondria from their mother, while males cannot transmit their mitochondria to subsequent generations.

Most human cells contain several hundred mitochondria. During mitotic cell division, the mitochondria of the dividing cell segregate in a purely random way to the two daughter cells. In each mitochondrion there are between about two and ten copies of the approximately 16.6 kb mitochondrial DNA molecule. Accordingly, although a single mitochondrial DNA molecule has only about 1/8000 as much DNA as an average sized chromosome, the total mitochondrial complement can account for up to about 0.5% of the DNA in a nucleated somatic cell. Although the mitochondrial DNA is principally double-stranded, a small section, the D loop, has a triple DNA strand structure due to the additional synthesis of a segment of mitochondrial DNA, 7S DNA (see *Figure 1.5*).

The 16.6 kb human mitochondrial DNA molecule contains 37 genes, 28 of which are encoded by the heavy (H) DNA strand, which is rich in guanines, and nine by the light (L) DNA strand (*Figure 1.5*). Of the 37 mitochondrial genes, 13 encode polypeptides which, together with the products of at least 50 nuclear genes, account for four of the five respiratory complexes, the multichain enzymes of oxidative phosphorylation, which are engaged in the production of ATP. The sub-units of the fifth respiratory complex, succinate–CoQ reductase, and all other mitochondrial proteins are encoded exclusively by the nuclear genome. The remaining 24 mitochondrial genes encode 22

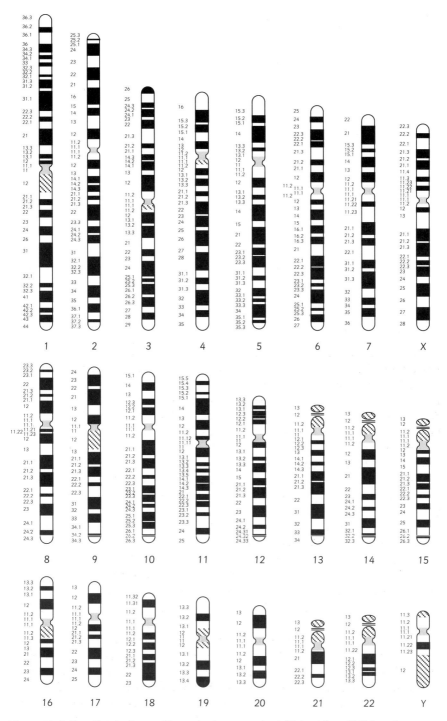

Figure 1.4: *Banding pattern of human chromosomes (G-banding, 550 band karyogram).*

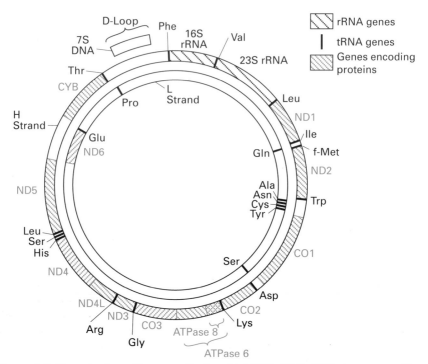

Figure 1.5: *Organization of human mitochondrial DNA. ND1–ND6: genes encoding NADH dehydrogenase sub-units 1–6. CO1–CO3: genes encoding cytochrome C oxidase sub-units 1–3. CYB–gene encoding cytochrome B.*

types of tRNA and two rRNA molecules which constitute part of the mitochondrial protein synthesis machinery; other components, for instance the aminoacyl tRNA synthetases, are encoded exclusively by nuclear genes.

Of the 22 types of tRNA encoded by mitochondria, eight can recognize families of four codons which differ only at the third base, and 14 recognize pairs of codons which are identical at the first two base positions and share either a purine or a pyrimidine at the third base. The remaining four codons, UAG, UAA, AGA, and AGG cannot be recognized by mitochondrial tRNA and act as stop codons (see *Table 1.2*).

Consequently, the genetic code employed to decipher mitochondrial-encoded mRNA on mitochondrial ribosomes differs from that used to decipher nuclear-encoded mRNA on cytoplasmic ribosomes. In addition, the mitochondrial and nuclear genomes differ in many other aspects of their organization and expression (*Table 1.3*).

1.3 Coding and non-coding DNA

Only a small fraction of the human genome (about 2–3%) is coding DNA, DNA sequences which directly specify a polypeptide or mature functional RNA product. The great majority of the genome is non-coding DNA (see *Figure 1.6*). At present, the bulk of extragenic DNA has no known function, either genetic or chromosomal, and accordingly has sometimes been considered to represent 'junk DNA'. Other extragenic non-coding DNA sequences have specific chromosomal functions.

Centromeres. Essential for ensuring proper disjunction of the chromosomes into daughter cells following cell division at meiosis and mitosis. Centromeric DNA predominantly comprises arrays of tandemly repeated DNA sequences, but the anticipated role of the latter in centromeric function has not been established (see Section 1.8.1).

Telomeres. These are required for ensuring complete replication of the DNA at the chromosome termini. In their absence, chromosomes become 'sticky' and will fuse with each other. In meiotic cells telomeres appear to be attached to the nuclear membrane and are the sites at which pairing of homologous chromosomes initiates. Telomeric DNA is composed of small arrays of tandemly repeated DNA (see Section 1.8.2).

Transcriptionally active DNA is generally marked by an altered chromatin structure which confers sensitivity to the enzyme DNase I, an endonuclease which nicks the individual strands of duplex DNA in a manner that is largely sequence-independent. Transcriptional activity is also often inversely correlated with the degree of methylation of cytosine residues. During DNA replication, transcriptionally active DNA is thought to be shielded from DNA methylases by transcription-associated protein factors. However, in other tissues where the same DNA is transcriptionally inactive, DNA methylases may gain access to the DNA and methylate it. The methylated DNA is subsequently bound by nuclear proteins such as MeCP-1, thereby limiting access to the DNA by transcription factors.

Table 1.2: *The genetic code in human cells*

AAA	Lys	ACA	Thr	AGA	Arg[N]/STOP[M]	AUA	Ile[N]/Met[M]
AAC	Asn	ACC	Thr	AGC	Ser	AUC	Ile
AAG	Lys	ACG	Thr	AGG	Arg[N]/STOP[M]	AUG	Met
AAU	Asn	ACU	Thr	AGU	Ser	AUU	Ile
CAA	Gln	CCA	Pro	CGA	Arg	CUA	Leu
CAC	His	CCC	Pro	CGC	Arg	CUC	Leu
CAG	Gln	CCG	Pro	CGG	Arg	CUG	Leu
CAU	His	CCU	Pro	CGU	Arg	CUU	Leu
GAA	Glu	GCA	Ala	GGA	Gly	GUA	Val
GAC	Asp	GCC	Ala	GGC	Gly	GUC	Val
GAG	Glu	GCG	Ala	GGG	Gly	GUG	Val
GAU	Asp	GCU	Ala	GGU	Gly	GUU	Val
UAA	STOP	UCA	Ser	UGA	STOP[N]/Trp[M]	UUA	Leu
UAC	Tyr	UCC	Ser	UGC	Cys	UUC	Phe
UAG	STOP	UCG	Ser	UGG	Trp	UUG	Leu
UAU	Tyr	UCU	Ser	UGU	Cys	UUU	Phe

[N,M]Alternative interpretations of nuclear and mitochondrial codons.

Table 1.3: *The human nuclear and mitochondrial genomes*

	Nuclear genome	Mitochondrial genome
Size	3000 Mb	16.6 kb
Number of different DNA molecules	23 (in XX) or 24 (in XY) cells, all linear	1 circular DNA molecule
Total number of DNA molecules per cell	23 in haploid cells; 46 in diploid cells	Several thousand
Associated protein	Several classes of histone and non-histone protein	Largely free of protein
Number of genes	50 000–100 000	37
Gene density	1/30–1/60 kb	1/0.45 kb
Repetitive DNA	Large fraction – see *Figure 1.6*	Very little
Transcription	The great bulk of genes are transcribed individually	Continuous transcription of multiple genes
Introns	Found in most genes	Absent
Percentage of coding DNA	2–3%	Approximately 95%
Codon usage	See *Table 1.2*	See *Table 1.2*
Recombination	At least once for each set of homologous chromosomes at meiosis	None
Inheritance	Mendelian for sequences on X and autosomes; paternal for sequences on Y	Exclusively maternal

1.3.1 Clustering of coding DNA and of non-coding DNA in defined chromosomal regions

As the nuclear chromosomes uncoil following cell division certain chromosomal regions continue to remain condensed throughout the life cycle of the cell. They appear as dark-staining areas, termed heterochromatin, and have been presumed to be genetically inactive. In contrast, the bulk of the chromatin is euchromatin which

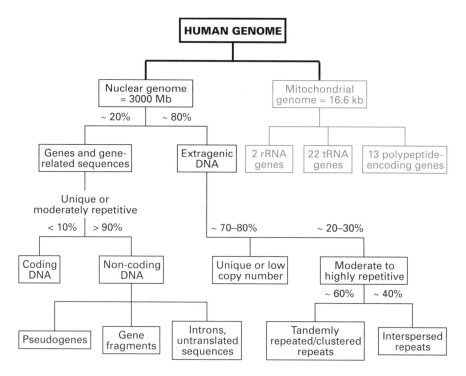

Figure 1.6: *Organization of the human genome.*

uncoils following cell division, becomes light-staining and includes active genes interspersed with non-transcribed DNA sequences. There are two classes of heterochromatin: facultative heterochromatin can be either genetically active or inactive as in the specialized case of X-chromosome inactivation (see Section 2.2); constitutive heterochromatin is always in the inactive state. Constitutive heterochromatic regions are composed almost entirely of certain repetitive DNA sequences which are found in and around the centromeres of all chromosomes, on the short arms of the acrocentric chromosomes 13, 14, 15, 21 and 22, in the secondary constrictions (light staining, apparently uncoiled chromosomal regions) of chromosomes 1, 9 and 16, and also on the bulk of the long arm of the Y chromosome. Although the vast majority, perhaps all, of the repetitive DNA sequences associated with constitutive heterochromatin are not transcribed into RNA, it is unlikely that there is complete absence of transcriptional activity within heterochromatic regions.

Although on average the human genome has about one gene per 30–60 kb of DNA, certain chromosomal regions such as the centromeres and telomeres are particularly deficient in genes. Other regions such as the Giemsa-negative bands are comparatively rich in genes. For example, the human major histocompatibility complex (human MHC or HLA complex) is located in a Giemsa-negative band, 6p21.3, and the gene density in a stretch of contiguous DNA spanning 680 kb in this region has been observed to be one per 19 kb. This cluster contains some examples of partially overlapping genes in which two genes are transcribed from overlapping DNA sequences but from the two different strands of the DNA helix. Other examples are known of whole genes being located entirely within other genes (see Section 1.5.4). However, in general, overlapping genes and genes within genes are rarely found in the human genome.

1.4 Regulation of gene expression

Most of the approximately 50 000–100 000 genes in the human genome specify polypeptides. A comparatively small minority are expressed to give a final product of RNA. This may be one of several types which are involved in the general process of gene expression. The remaining genes specify polypeptides. Of these, some are expressed at low levels in all somatic cells to give products which are required for general cell function (housekeeping genes). The expression of other genes, however, is tissue-specific: for example, the β-globin gene is expressed in certain blood cells but not in muscle whereas the reverse is true for the dystrophin gene. Often tissue-specific expression of a gene is simply controlled at the level of transcription (Section 1.4.2) resulting in the synthesis of an identical gene product in one or a small number of tissues. However, in many cases tissue-specific alternative forms of the same protein (isoforms), or of the same enzyme (isozymes), can co-exist in the same individual, as can isozymes specific for different subcellular compartments. Such alternative protein forms can be synthesized by different but highly related genes (Section 4.4.3) or by differential use of alternative promoters, splice signals or polyadenylation sites (Section 2.8).

1.4.1 Classes of RNA polymerase and their products

In addition to the mitochondrial ribosomal and tRNA species, there are several classes of nuclear genes whose ultimate products are RNA molecules. The bulk of the ribosomal RNA found on cytoplasmic ribosomes, including the 28S rRNA and 5.8S rRNA components of the large ribosomal sub-unit, and the 18S rRNA component of the small ribosomal sub-unit, are transcribed by RNA polymerase I in the nucleolus. In this specialized case, the genes specifying 28S, 5.8S and 18S rRNA are clustered on a single 13 kb length of DNA (transcription unit) which is transcribed into a single large precursor RNA molecule. This primary transcript subsequently undergoes various RNA processing steps which ultimately generate the individual 28S, 5.8S and 18S rRNA species. RNA polymerase III, which is located in the extranucleolar region of the nucleus, is employed to transcribe a variety of genes, including those specifying the various transfer RNA species, 5S rRNA (a constituent of the large sub-unit of cytoplasmic ribosomes) and a diverse series of small RNA species, including most of the classes of SnRNA (small nuclear RNA) which are involved in RNA processing.

The vast majority of cellular genes are individually transcribed by RNA polymerase II. The transcriptional start point is dictated by a promoter sequence, typically located within a few hundred bp upstream of the gene. Transcription normally terminates shortly after the RNA polymerase encounters a special sequence, usually a variant of the consensus sequence AATAAA, which is involved in a subsequent processing step in which the RNA is polyadenylated (see Section 1.5.2).

Table 1.4: *Examples of* cis-*acting and* trans-*acting transcriptional elements*

Cis element	DNA sequence is identical to, or a variant of	Associated trans-acting factors	Comments
GGGCGG box	GGGCGG	Sp1	Sp1 factor is ubiquitous
TATA box	TATAAA	TFIID	TFIIA binds to the TFIID–TATA box complex to stabilize it
CCAAT box	CCAAT	Many, e.g. C/EBP, CTF/NF1	Large family of *trans*-acting factors
CAT box	CAT	TEF2	
TRE (TPA response element)	GTGAGTA/CA	AP-1 family, e.g. *JUN/FOS*	Large family of *trans*-acting factors
CRE (cAMP response element)	GTGACGTA/CAA/G	CREB/ATF family, e.g. ATF-1	Genes activated in response to cAMP
PE element	GTTAATNATTAAC	HNF-1	HNF-1 is liver-specific
Octa element	ATGCAAAT	OTF-1 OTF-2	OTF-1 is found in many cell types; OTF-2 is mostly restricted to lymphocytes, and is the limiting factor in expressing immunoglobulin genes
GATA element	GATA	GATA-1 (= NF-E1)	GATA-1 is erythroid-specific and present at all developmental stages

1.4.2 Transcriptional control of gene expression

Gene expression in human cells is mostly regulated at the level of transcription [5]. *Cis*-acting transcription factors are elements of defined DNA sequence or structure which are located on the same DNA strand as the genes they regulate and in the vicinity of the genes (*Table 1.4*). They do not encode any product as such but are thought to influence transcription directly. They often act as binding sites for *trans*-acting transcription factors. The latter are protein products encoded by other genes, which influence the transcriptional activity of genes by binding to specific DNA regions in the vicinity of these genes. Certain *trans*-acting transcription factors are known to be confined to a limited set of cell types (*Table 1.4*), and are therefore expected to make a major contribution to tissue-specific gene expression (*Table 1.5*). The *cis*-acting sequence elements are often organized in clusters which typically span 100–300 bp, including the following.

Promoters. Combinations of various short sequence elements in the DNA region immediately upstream of genes. They govern the start position and general level of transcription.

Enhancers/silencers. Combinations of sequence elements which stimulate the transcription of genes (enhancers) or suppress transcription (silencers). Although they may be located near or even within the gene whose expression they influence, they are often located some distance away. Their precise position or orientation is not critical for their function.

Table 1.5: *Different levels of selective gene expression*

Selective expression mechanism	Examples
Activation of expression of selected genes by inherent tissue-specific transcription factors	Gene activation involving: GATA-1 (erythroid-specific) HNF-1 (liver-specific) TCF-1 (T lymphocyte-specific)
Activation of expression of selected genes by action of extrinsic factors on inducible promoters/enhancers	cAMP-mediated activation of transcription
Tissue-specific differential transcription or alternative RNA processing of specific genes	Differential use of alternative promoters, splice signals, polyadenylation sites, etc. in different cell types (Section 2.8)
Activation of expression of selected genes by stimulation of response elements in mRNA	Activation of iron-response elements in ferritin and transferrin mRNA (Section 1.4.3)
Developmental stage-specific expression	Hemoglobin switching (*Figure 1.10*)
DNA rearrangements to generate cell-specific gene expression	Production of cell-specific immunoglobulins and T-cell receptors in B and T lymphocytes, respectively (Section 2.7.3)

1.4.3 Translational control of gene expression

In addition to transcriptional control of gene expression, the expression of some human genes is known to be controlled at the translational level. For example, increased iron levels stimulate the synthesis of the iron-binding protein, ferritin, without any corresponding increase in ferritin mRNA. In this case, the stimulated synthesis occurs at the translational level. The effect appears to be brought about by an iron-responsive *trans*-acting factor which binds to regulatory elements located in the ferritin mRNA [6].

1.5 Expression of polypeptide-encoding genes

1.5.1 Transcription and regulation of gene expression

Genes which are very actively transcribed, either at a specific stage in the cell cycle (e.g. histones) or in specific cell types (e.g. β-globin), always have a TATA box in their promoter. This element, often TATAAA or a variant, normally occurs at a position about 30 bp upstream (−30) from the transcriptional start site (*Figure 1.7*). However, TATA boxes are absent from the promoters of many other genes, including housekeeping genes. Instead, the latter often have GC-rich sequence elements, especially variants of the consensus sequence GGGCGG (*Table 1.4*). Other common promoter elements include the CAAT box, usually at about −80, which is often the strongest determinant of promoter efficiency.

A number of other regulatory elements are known to be required for efficient expression of many human genes, and sequence elements which stimulate or suppress transcription have been identified in the immediate vicinity of some genes, and occasionally intragenically. For example, the human platelet-derived growth factor-B gene (*PDGFB*) is known to be regulated by both positive and negative tissue-specific regulatory elements located in the first intron [7]. In the case of the human β-globin gene a number of different regulatory elements have been identified (*Figure 1.7*). Notwithstanding the presence of intragenic and 3' flanking enhancers, the major role in stimulation of transcription *in vivo* is played by a locus control region (LCR), located approximately 50–60 kb upstream of the β-globin gene; deletion of this LCR results in inactivation of the β-globin gene and contributes to β-thalassemia. The β-globin cluster LCR includes four short (200–400 bp) regulatory regions (HS-1 to HS-4) which are hypersensitive to DNase I in cells of the erythroid lineage and which are developmentally stable. Individual HS sites contain multiple *cis* elements, including recognition sequences for both ubiquitous transcription factors and erythroid-specific transcription factors such as GATA-1.

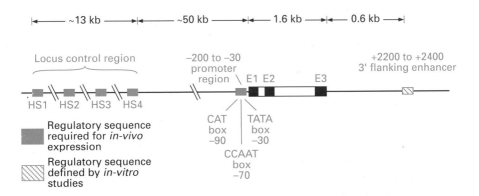

Figure 1.7: *Important regulatory sequences required for expression of the human β-globin gene. E1–E3: exons in the β-globin gene.*

1.5.2 Post-transcriptional processing

Usually, the coding sequences of polypeptide-encoding genes are split into segments (exons) which are separated by non-coding intervening sequences (introns). The primary RNA transcripts from such genes normally represent an RNA copy of an individual gene sequence, including sequences complementary to both exons and introns (*Figure 1.8*). Subsequent RNA processing events to generate mRNA include the following.

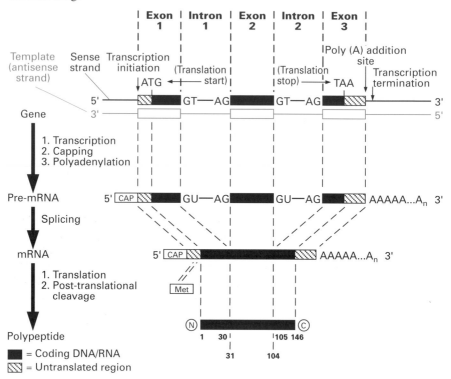

Figure 1.8: *Expression of the human β-globin gene.*

Capping. This involves blocking the 5' end of the mRNA by the attachment, through a 5'–5' triphosphate linkage, of a specialized nucleotide containing 7-methylguanine.

Polyadenylation. About 200 adenylate residues are sequentially added enzymatically to form a poly(A) tail at the 3' end of the RNA transcribed from all genes except the histone genes.

Splicing. The RNA is cleaved so that sequences corresponding to introns are excised and discarded. The remaining RNA segments corresponding to exons are spliced together. The location of the splice sites is dictated by short sequences at the intron/exon boundaries in the gene. A GT sequence which is found at the 5' end of each intron (and becomes GU in the transcribed RNA) and an AG sequence at the 3' end of each intron are normally essential, but not sufficient, for correct splicing.

1.5.3 Translation and post-translational processing

In the mRNA molecule the sequence which specifies polypeptide is flanked by 5' and 3' untranslated sequences. Such sequences, which were originally copied from sequences within the 5' and 3' terminal exons, do not specify polypeptide but are required, like the 5' cap and 3' poly(A) tail, to assist in binding and stabilizing the mRNA on the ribosomes. The remaining mRNA sequence is translated into a polypeptide in which the individual amino acids are specified by successive groups of three nucleotides (codons), normally beginning with an N-terminal amino acid of methionine specified by the codon AUG. Polypeptide synthesis is terminated upon encountering a stop codon: UAA, UAG or UGA in the case of nuclear-encoded mRNA; UAA, UAG, AGA or AGG in the case of mitochondrial-encoded mRNA (*Table 1.2*). Some polypeptides are then processed by hydroxylation of specific amino acids, notably proline and lysine, or by sequential addition of sugars to specific amino acids to form carbohydrate side-chains (glycosylation).

The maturation of many human polypeptides, including plasma proteins, polypeptide hormones, neuropeptides and growth factors, requires the post-translational proteolytic cleavage of a precursor polypeptide with minor peptide cleavage products being discarded. In some cases, cleavage of a single precursor polypeptide may generate more than one functional polypeptide chain as in the synthesis of insulin, serum complement factors and certain other proteins. All polypeptides whose ultimate destination requires passage through cell membranes (e.g. secreted polypeptides, mitochondrial proteins synthesized in the cytoplasm, etc.) are also synthesized initially as precursor polypeptides. Such precursors have an additional signal sequence (sometimes called a leader sequence) of about 16–30 amino acids at the N-terminal end which acts as a recognition signal for transport across cellular membranes. Thereafter the signal peptide is cleaved from the main polypeptide and degraded. Finally, other polypeptides often have the initial 5' methionine removed by proteolytic cleavage, as in the case of β-globin (see *Figure 1.8*).

1.5.4 Organization of polypeptide-encoding genes

There is extensive variation in the size and intron/exon organization of human genes (*Table 1.6*). Although there is some degree of correlation between the size of a gene and the size of its product, and also between the size of a gene and the number of exons, there are striking anomalies. The dystrophin gene is about 50 times larger than the apolipoprotein B gene but encodes a significantly smaller product, and the 18 kb type I collagen gene is less than one-tenth of the size of the factor VIII gene but has 52 exons. Generally, however, the average size of exon shows less variation than the average intron size, and there is a strong tendency for large genes to have very large introns (e.g. intron 44 of the dystrophin gene is about 170 kb long). Large introns occasionally contain whole small genes which are transcribed from the opposite DNA strand to that used to express the larger gene. For example, the clotting factor VIII gene contains a single small intronless gene within one of its introns and one large intron of the NF1 (neurofibromatosis type I) gene contains three small genes, each with two exons (*Figure 1.9*). Presently, the precise significance of introns in human genes is uncertain (see Section 2.4).

Table 1.6: *Size and exon/intron organization of human genes*

Gene	Size of gene (kb)	Size of polypeptides (number of amino acids)[a]	Number of exons	Amount of intron (%)	Average size of exon (bp)	Average size of intron (kb)
tRNA[Tyr]	0.1	n/a	2	18	50	0.02
Histone H4	0.4	102	1	0	—	—
α-interferon	0.9	23(S) + 166	1	0	—	—
Insulin	1.4	24(S), 30(B) 31(C), 21(A)	3	67	155	0.48
β-globin	1.6	146	3	62	212	0.49
Class I HLA	3.5	24(S) + 340	8	54	187	0.26
Serum albumin	18	18(S) + 585	14	88	137	1.1
Complement C3	41	22(S), 645(A) + 992(B)	29	88	122	0.9
Apolipoprotein B	43	4536 (2152[b])	18	67	783	1.7
LDL receptor	45	21(S) + 839	13	89	394	3.3
Phenylalanine hydroxylase	90	451	26	97	96	3.5
Factor VIII	186	19(S) + 2332	24	97	375	7.8
Cystic fibrosis trans-membrane regulator	250	1480	27	97.6	227	9.1
Dystrophin	2300	3700	77	99.4	≈180	≈30

[a]Mature gene products are underlined. A = A chain, B = B chain, C = connecting peptide, S = signal peptide.
[b]Major size in intestinal cells due to differential processing (see Section 2.8).

1.6 Repetitive DNA

In addition to the presence of two alleles at each locus in the diploid genome, approximately 30–40% of the nuclear genome in both haploid and diploid cells is composed of sets of closely related non-allelic DNA sequences (repetitive DNA). Within the considerable variety of different repetitive DNA sequences in the haploid genome are sequence families whose individual members include functional genes (multigene families), and also many examples of non-genic repetitive DNA sequence families. In each case the operational definition of a DNA sequence family is the typically high level of DNA sequence similarity (sequence homology) between individual repeat unit members of the family. This can be demonstrated by direct DNA sequencing of different repeats, or more conveniently, by DNA hybridization assays (see Section 3.1.4). When two members of a repetitive DNA sequence family exhibit a high degree of sequence homology, a recent common evolutionary origin is indicated. As detailed in the following sections, DNA sequence families show considerable variation in the number of different repeat-unit members in the family, the size of the repeating unit, chromosomal localization, mode of repetition, and capacity for expression.

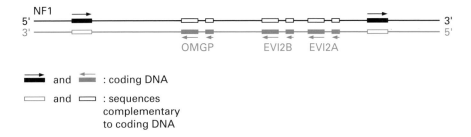

Figure 1.9: Genes within genes: the human NF1 gene contains three internal genes which are located between two exons [8].

1.7 Multigene families

A large percentage of actively expressed human genes are members of families of DNA sequences which show a high degree of sequence similarity but which can vary in their function. Two types of sequence organization are seen: (a) tandem repetition or clustering of individual members of the family; (b) interspersion of family members.

1.7.1 RNA encoding gene families

As indicated in *Table 1.7*, genes whose final expression product is an RNA molecule belong to multigene families which are among the most repetitious in the genome. The genes which encode 28S, 5.8S and 18S rRNA are included in repeated units of DNA approximately 45 kb long, comprising the common 13 kb transcription unit (see Section 1.4.1) and an adjacent non-transcribed spacer DNA. Arrays of about 50–70 such tandemly repeated units are located on the short arms of each of the five acrocentric chromosomes. The other major RNA constituent of ribosomes, 5S rRNA, is transcribed from genes which constitute part of a very large gene family which is clustered on the long arm of chromosome 1. Including spacer units, the rRNA gene families account for about 0.4% of the total DNA in the genome.

Other RNA gene families include the different but closely related transfer RNA gene families. The genes which specify each of the more than 50 different types of tRNA are members of multigene families which individually have copy numbers between 10 and 100 and collectively have a copy number of about 1600. Genes encoding the different classes of SnRNA molecule also belong to large gene families. One common characteristic of each of these gene families is that they include a considerable number of pseudogenes, family members which may show extremely high levels of sequence homology to the functional gene members but which are defective in gene expression as a consequence of the acquisition of deleterious mutations. The functional significance of pseudogenes, if any, is not clear, and they most probably represent almost inevitable evolutionary by-products of the genetic mechanisms that give rise to duplicated genes (see Sections 2.3 and 2.7.7). Additionally, several members of the gene families are gene fragments, truncated genes which lack either the 5' or 3' ends of the normal genes.

Table 1.7: Examples of human multigene families

Family	Copy number	Organization	Chromosome location
Complement C4	2	On tandem compound repeats ≈ 30 kb in length	6p21.3
Aldolase	5	Interspersed, three functional genes and two pseudogenes	3, 9q, 10, 16q, 17
Growth hormone cluster	5	Clustered within 67 kb, one pseudogene	17q22-24
Ferritin heavy chain	>15	Interspersed, most are pseudogenes, but at least one active gene on 11	Many
Glyceraldehyde 3-phosphate dehydrogenase	>18	Interspersed, one functional gene on 12p, many pseudogenes	Many
Class I HLA heavy chain	~20	Clustered over 2 Mb; at least four are expressed, but many are pseudogenes and gene fragments	6p21.3
Actin	>20	Interspersed, four functional genes and many pseudogenes	Many
β-tubulin	20–30	Interspersed, three functional genes and many pseudogenes	Many
Histone	>100	Clustering at a few locations, especially compound cluster on chromosome 1p21	1p21, 6, 12q
28S + 5.8S + 18S rRNA	300+	Five arrays of tandem, ≈ 40 kb, repeats	13,14,15, 21, 22
tRNA	1600	Many different subfamilies dispersed throughout genome	Many

1.7.2 Polypeptide encoding gene families

Genes which encode polypeptides are often members of multigene families whose copy number can vary widely. In some cases, such as the genes which encode an individual type of histone molecule, the sequences of the different members of a high copy number gene family are nearly identical. In other multigene families, accumulated sequence divergence between the family members may result in divergence of function. In DNA sequence families which have members in multiple clusters, there is generally a larger degree of repeat sequence identity within a cluster than between clusters. This difference may reflect differences in the rate of intra- and interchromosomal DNA sequence exchange.

Functional divergence between members of a multigene family may extend to the loss of function by some family members. For example, clustered multigene families often include conventional pseudogenes, gene copies which, although they retain sequences homologous to exons, introns and immediate flanking sequences of a functional gene, are nevertheless functionless (*Figure 1.10* and *Table 1.7*). Other gene members may be functionless gene fragments following loss of one or more exons, probably as a result of unequal cross-over or of unequal sister chromatid exchange (see Section 2.7).

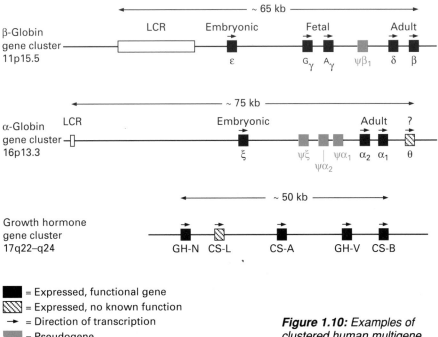

Figure 1.10: *Examples of clustered human multigene families.*

= Expressed, functional gene
= Expressed, no known function
→ = Direction of transcription
= Pseudogene
LCR = Locus control region

Less extreme forms of functional divergence include gene-specific differences in the developmental timing and tissue-specificity of gene expression. The regulation of individual gene expression in a gene family may accordingly be complex. For example, in both the α- and β-globin gene clusters individual genes become active at different developmental stages (hemoglobin switching) and there is a strong correlation between the linear order of the genes and the time at which they become active in development (*Figure 1.10*). Recently, it has become apparent that a dominant control region, the LCR exists in both gene clusters. Such cluster-specific LCRs are thought to organize the cluster into an active chromatin domain and to act as an enhancer of globin gene transcription. Developmental stage-specific switching in globin gene expression is then thought to be accomplished by competition between the globin genes for interaction with their respective LCR and stage-specific activation of gene-specific silencer elements. For example, transcription of the ε-globin gene is preferentially stimulated by the neighboring LCR at the embryonic stage. In the fetus, however, ε-globin expression is suppressed following activation of a silencer and γ-globin expression becomes dominant (*Figure 1.11*).

In addition to clustered gene families there are numerous examples of interspersed gene families which show little or no evidence of clustering (*Table 1.7*). Some of the interspersed gene families contain only a few members, all or most of which are functional, and which probably arose as a result of ancient genome or gene duplication events (see Sections 2.2 and 2.3). However, the majority of the interspersed gene families contain one or a small number of active genes and a large number of processed pseudogenes. The latter differ from the conventional pseudogenes seen in clustered gene families in that they consistently lack any sequences corresponding to the introns or promoters of the functional gene members. Processed pseudogenes are thought to arise by an RNA-mediated DNA transposition mechanism (Section 2.7.7).

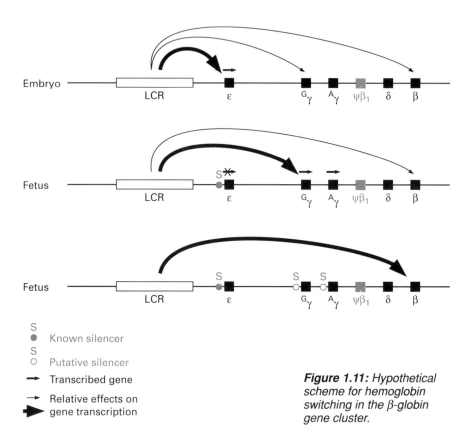

S
● Known silencer

S
○ Putative silencer

→ Transcribed gene

→ Relative effects on
▶ gene transcription

Figure 1.11: *Hypothetical scheme for hemoglobin switching in the β-globin gene cluster.*

An extension of the concept of a gene family is that of a gene superfamily, such as the immunoglobulin superfamily, which comprises several families of genes of distinct, but related, functions. The individual members of superfamilies show, on average, a much lower degree of sequence identity than members of a gene family, reflecting an earlier evolutionary divergence of their members.

1.8 Extragenic repeated DNA sequences

Repeated DNA sequence families which do not include functional gene members are composed of arrays of tandem repeats, or of individual repeat units interspersed with other DNA sequences (*Table 1.8*). The former category may be sub-divided according to the average size of the arrays of tandem repeats into satellite DNA, minisatellite DNA and microsatellite DNA [9].

1.8.1 Satellite DNA

Satellite DNA comprises large arrays of tandemly repeated DNA which usually fall within the 100 kb to several megabases range, and consist of simple or moderately complex repeat units. Repeated DNA of this type is not transcribed and accounts for the bulk of the heterochromatic regions of the genome (see Section 1.3.1). The base composition, and therefore density, of such DNA regions is dictated by the base composition of their constituent short repeat units and may diverge substantially from the overall base composition of bulk cellular DNA. Consequently, it is possible to isolate classical satellite DNA species by buoyant density-gradient centrifugation where they appear as minor components (classical satellite DNA) of different buoyant density from bulk DNA. Three major human satellite DNA species have been resolved on Ag–CsSO$_4$ gradients: I – 1.687 g cm^{-3}; II – 1.693 g cm^{-3}; III – 1.697 g cm^{-3}. Each satellite class includes a number of different tandemly repeated DNA sequence families (satellite subfamilies), some of which are shared between different classes. DNA sequence analysis has revealed that some of the repetitive DNA families in the satellites are based on very simple repeat units. For example, both satellite II and satellite III contain sequence arrays which are based on tandem repetition of the

Table 1.8: Major extragenic repetitive DNA classes

Class	Size of repeat unit (bp)	Total number of copies of repeat units	Major chromosomal location(s)
Tandem repeats			
Satellite DNA			Heterochromatin
Simple sequence	5–25[a]	?	of 1q, 9q, 16q, Yq
Alpha (alphoid DNA)	171[a]	8 x 10^5	of centromeres
Beta (*Sau*3A family)	68[a]	5 x 10^4	of 9, 13, 14, 15, 21, 22
Minisatellite DNA			
Telomeric family	6	2–3 x 10^4	Telomeres
Hypervariable family	9–64	3 x 10^4	All chromosomes, often near telomeres
Microsatellite DNA			
$(A)_n/(T)_n$	1	10^7	All chromosomes
$(CA)_n/(TG)_n$	2	7 x 10^6	All chromosomes
$(CT)_n/(AG)_n$	2	3 x 10^6	All chromosomes
Interspersed repeats			
Alu family	250[b]	7 x 10^5	Euchromatin, Giemsa-positive bands
Kpn (L1) family	1300[b]	6 x 10^4	Euchromatin, Giemsa-negative bands

[a]Higher order periodicities may also be observed.
[b]Approximate average size in genome.

sequence ATTCC. Additionally, restriction mapping (Section 3.2.1) has revealed satellite subfamilies which show additional higher order repeat units superimposed on the small basic repeat units. Such subfamilies are thought to arise as a result of subsequent amplification of a unit which is larger than the initial basic repeat unit and contains some diverged units (*Figure 1.12*).

Certain satellite DNA species cannot be resolved by density-gradient centrifugation but can be identified following digestion with a restriction nuclease which typically has a single recognition site in the basic repeat unit. Alpha satellite (or alphoid DNA) constitutes the bulk of the centromeric heterochromatin on all of the chromosomes. It is characterized by tandem repeats of a basic mean length of 171 bp, although higher order units are also seen. The sequence divergence between individual members of the alphoid DNA family can be so high that it is possible to isolate repeat units which will hybridize under stringent conditions to specific chromosomes.

Presently, the extent to which satellite DNA can be considered 'junk DNA' is not known. The centromeric DNA of human chromosomes largely consists of various families of satellite DNA. Of these only the alpha satellite is known to be present on all chromosomes, although sequence divergence between the repeats has led to chromosome-specific sub-sets. Although the 171 kb repeat unit of alpha satellite DNA often contains a binding site for a specific centromere protein, CENP-B, there is presently no compelling evidence to suggest that centromere function depends on this association, or even on the presence of alpha satellite DNA.

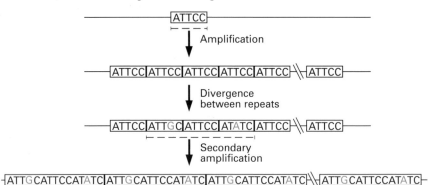

Figure 1.12: *Formation of higher order repeat units in simple sequence satellite DNA.*

1.8.2 Minisatellite DNA

Minisatellite DNA comprises a collection of moderately sized arrays of tandemly repeated DNA sequences which are dispersed throughout the nuclear genome. They include a family of hypervariable minisatellite DNA sequences which are organized in over 1000 arrays of short (from 0.1 to 20 kb long) tandem repeats. The repeat units in different hypervariable arrays vary considerably in size (*Table 1.7*), but share a common core sequence, GGGCAGGAXG (where X = any nucleotide), which is similar in size and in G content to the *chi* sequence, a signal for generalized recombination in *Escherichia coli*. While many of the arrays are found near the telomeres, several hypervariable minisatellite DNA sequences occur at other chromosomal locations.

Another major family of minisatellite DNA sequences is found at the termini of chromosomes (telomeres). The principal constituent of telomeric DNA is 10–15 kb of tandem hexanucleotide repeat units, especially TTAGGG, which are added by a specialized enzyme, telomerase. By acting as buffers to protect the ends of the chromosomes from degradation and loss during DNA replication, these simple repeats are directly responsible for telomere function.

1.8.3 Microsatellite DNA

Microsatellite DNA families include small arrays of tandem repeats which are simple in sequence (generally 1–4 bp) and are interspersed throughout the genome. Of the mononucleotide repeats, runs of A and of T are very common and together account for about 10 Mb, or 0.3% of the nuclear genome. In contrast, runs of G and of C are very much rarer. In the case of dinucleotide repeats, arrays of CA repeats (TG repeats on the complementary strand) are very common, accounting for 0.5% of the genome, and are often highly polymorphic. CT/AG repeats are also common, occurring on average once every 50 kb and accounting for 0.2% of the genome, but CG/GC repeats are very rare as a consequence of selective methylation and subsequent deamination of C residues which are flanked at their 3' end by a G residue (i.e. CpG – see Section 2.6). Trinucleotide and tetranucleotide tandem repeats are comparatively rare.

The significance of microsatellite DNA is not known. Alternating purine–pyrimidine repeats such as tandem repeats of the dinucleotide pair CA/TG are capable of adopting an altered DNA conformation, Z-DNA, *in vitro*, but there is little evidence that they do so *in vivo*. Although microsatellite DNA has generally been identified in intergenic DNA or within the introns of genes, a few rare examples have been recorded within the coding sequences of genes.

1.8.4 Highly repeated interspersed DNA

There are two major repetitive DNA families in this class. The *Alu* repeat family is the most conspicuous human example of a class of mammalian short interspersed nuclear elements, SINEs. The *Kpn* repeat family (also called the Line-1 or *L1* family) is an example of a mammalian LINE (long interspersed nuclear element) [10,11]. On the basis of the high copy number of these repeats (*Table 1.8*), an *Alu* repeat would be expected once every 4 kb on average, and a *Kpn* repeat once every 50 kb, and inspection of established DNA sequences generally confirms these predictions.

The *Alu* family repeats have a relatively high GC content, and although dispersed mainly throughout the euchromatic regions of the genome, appear to show preferential localization in the pale G-bands (Giemsa-negative) of metaphase chromosomes. While they are conspicuously absent from coding sequences, they are often found in non-coding intragenic locations, notably in introns and occasionally in untranslated sequences. Consequently, they are often represented in the primary transcript RNA from genes, and occasionally in mRNA. Many of the *Alu* repeats appear to be capable of being transcribed by RNA polymerase III into short RNA species which turn over rapidly. Present data suggest that the *Alu* repeats have evolved by RNA-mediated DNA transposition (see Section 2.7.7) from 7SL RNA, a small RNA species which constitutes part of the signal recognition particle used to transport secreted proteins across the endoplasmic reticulum. Although the function of *Alu* sequences is unknown, they have been considered to promote unequal recombination (Section 2.7.2).

The *Kpn* family repeats are heterogeneous in length, sharing a common 3' end, terminating in an A-rich tract, but differing widely in the amount of 5' end sequence, such that full-length sequences of 6–7 kb are relatively rare. They are primarily located in euchromatic regions but show an inverse relationship with *Alu* repeats by appearing to be preferentially located in the dark G-bands (Giemsa-positive) of metaphase chromosomes. Like the *Alu* repeats they are absent from coding sequences. Although they may be found in intragenic non-coding sequences and consequently are represented in primary transcript RNA, they are virtually absent from mRNA. One component of the *Kpn* repeat shows strong sequence similarity to known transposon genes which encode reverse transcriptase, and, like the *Alu* repeat family, the *Kpn* family is thought to contain some actively transposing members (Section 2.7.7). However, the function, if any, of the *Kpn* repeats is unknown.

References

1. Stephens, J.C., Cavanaugh, M.L., Gradie, M.I., Mador, M.L. and Kidd, K. (1990) *Science,* **250**, 237.
2. Manuelidis, L. (1990) *Science,* **250**, 1533.
3. Bernardi, G. (1989) *Ann. Rev. Genet.,* **23**, 637.
4. Anderson, S., Bankier, A.T., Barrell, B.G. *et al.* (1981) *Nature*, **290**, 457.
5. Johnson P.F. and McKnight, S.L. (1989) *Ann. Rev. Biochem.,* **58**, 799.
6. Hentze, M.W., Caughman, S.W., Ronault, T.A. *et al.* (1987) *Science*, **238**, 1570.
7. Franklin, G.C., Donovan, M., Adam, G.I.R., Holmgren, L., Pfeifer-Ohlsson, S. and Ohlsson, R. (1991) *EMBO J.*, **10**, 1365.
8. Cawthon, R.M., Andersen, L.B., Buchberg, A.M. *et al.* (1991) *Genomics,* **9**, 446.
9. Vogt, P. (1990) *Hum. Genet.*, **84**, 301.
10. Hutchison, C.A. III, Hardies, S.C., Loeb, D.D., Shehee, W.R. and Edgell, M.H. (1989) in *Mobile DNA* (D.E. Berg and M.M. Howe, eds). American Society for Microbiology, Washington D.C., p. 593.
11. Moyzis, R.K., Torney, D.C., Meyne, J., Buckingham, J.M., Wu, J.R., Burks, C., Sirotkin, K.M. and Goad, W.B. (1989) *Genomics*, **4**, 273.

Further reading

Darnell, J., Lodish, H. and Baltimore, D. (1990) *Molecular Cell Biology*, 2nd edn. Scientific American Books, New York.

Human Gene Mapping 10 (1989) Tenth international workshop on human gene mapping. *Cytogenet. Cell Genet.*, **51**.

Kao, F.-T. (1985) Human genome structure. *Int. Rev. Cytol.*, **96**, 51.

Lewin, B. (1990) *Genes IV.* Oxford University Press, Oxford.

Singer, M. and Berg, P. (1991) *Genes and Genomes.* University Science Books, California.

Various authors (1986) *The Molecular Biology of Homo Sapiens.* Cold Spring Harbor Symposiums in Quantitative Biology, Volume 51.

2
EVOLUTION AND POLYMORPHISM OF THE HUMAN GENOME

In many ways the human genome is a typical higher eukaryote genome. The genome size is large, a parameter which, with some notable exceptions, parallels the complexity of the organism (*Table 2.1*). It contains a large fraction of non-coding DNA sequence whose significance is mostly obscure. In addition, a high proportion of both the coding DNA and non-coding DNA is repetitive. By comparison, the genomes of prokaryotes are small, with little non-coding DNA or repeated DNA sequences.

Table 2.1: *Interspecific variation in chromosome number and genome size*

Species	Haploid chromosome number	Haploid genome size (Mb)
Saccharomyces cerevisiae (yeast)	16	14
Dictyostelium discoideum (slime mould)	7	70
Caenorhabditis elegans (nematode)	11/12	100
Drosophila melanogaster (fruit fly)	4	170
Gallus domesticus (chicken)	39	1200
Mus musculus (mouse)	20	3000
Xenopus laevis (toad)	18	3000
Homo sapiens (human)	23	3000
Zea mays (corn)	10	5000
Allium cepa (onion)	8	15 000

2.1 Origin of the nuclear and mitochondrial genomes

Many features of the human mitochondrial genome are reminiscent of those of prokaryotic genomes, including the structure of the ribosomal genes and the general lack of extragenic and intragenic non-coding DNA sequences. These and other similarities between mitochondria and bacteria have suggested that mitochondria were generated as a result of endocytosis by anerobic eukaryotic precursor cells of a bacterium, probably a purple photosynthetic eubacterium [1]. By subverting the bacterial oxidative phosphorylation system the eukaryotic precursor cells were able to promote their own rapid growth and evolution in an oxygen-containing atmosphere.

As the size of the human mitochondrial genome is only a small fraction of that of a typical bacterial genome, it is probable that many of the genes donated by the eubacterial endosymbiont were gained by the nuclear genome and that the protein products of these previous eubacterial genes are imported into the mitochondria. The mitochondrial genome presumably retained only a very small number of the eubacterial genes, including the core rRNA genes and some tRNA genes. The unique genetic code of mitochondria may have evolved as a response to the limited coding capacity of this genome. As the human mitochondrial genome directs the synthesis of only 13 different types of polypeptide, the possibility of disastrous consequences from slight altering of the otherwise universal genetic code is minimized. Instead, it is likely that the codons which have been altered (see *Table 1.2*) have not been used extensively in locations where amino acid substitutions would have been deleterious.

Although mitochondrial DNA is evolving more rapidly than nuclear DNA (Section 2.6), the absence of recombination and the strict maternal inheritance has facilitated studies on human evolution. Analysis of the mitochondrial DNA of 147 individuals representing five human geographic populations has prompted the remarkable suggestion that they, and perhaps all contemporary human mitochondria, are descended from a single woman who lived about 200 000 years ago [2].

2.2 Evolution of genome size and chromosome organization

The relative complexity of the human genome in terms of size and number of different chromosomes is presumed to reflect a succession of ancestral DNA duplications and rearrangements. Genome duplication is an effective way of increasing genome size and is responsible for the extensive polyploidy of many flowering plants. Possibly, genome duplication events were involved in the evolution of the human genome to create transient tetraploid genomes which were restored to the diploid state after subsequent chromosome divergence by translocations and inversions. In the human genome certain specific pairs of non-homologous chromosomes have been claimed to show significant similarity between their sizes and banding patterns, indicative of a recent common origin by genome duplication [3]. Additionally, there is some evidence that such pairs of putative homeologous chromosomes often exhibit closely related pairs of non-allelic genes. However, the evidence for proposed ancestral genome duplications has been obscured by subsequent chromosome rearrangements. Probably ancestral genome duplication events were extremely rare compared to the slow accumulation of DNA sequences by chromosomal translocations and by copying mechanisms such as tandem DNA duplication, DNA amplification and DNA transposition (see below).

The organization and banding patterns of human chromosomes are very similar to those of other primates. The major difference is that other primates have 23 different autosomes; two ancestral chromosomes (corresponding to human 2p and 2q) appear to have fused in the human lineage but not in those leading to the other primates. Other minor differences are mostly accounted for by chromosome inversions and variations in constitutive heterochromatin; differences due to chromosome translocations are relatively rare. On first examination, mouse and human chromosomes would appear to have very different organizations (the mouse has 20 pairs of acrocentric chromosomes whereas there are 23 pairs of human chromosomes, most of which are metacentric or sub-metacentric). However, comparison of high resolution mouse and human chromosome maps also reveals considerable sharing of cytogenetic banding patterns over relatively small chromosomal regions. Small chromosomal regions appear to have been conserved, therefore, over comparatively long evolutionary time-scales (see also Section 4.4.5).

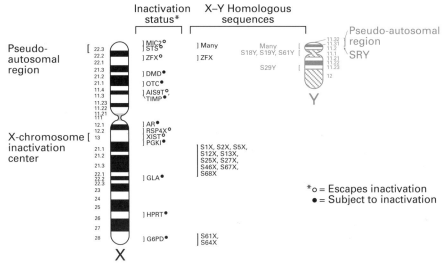

Figure 2.1: *The human sex chromosomes.*

Unlike autosomal homologous chromosome pairs, the X and Y chromosomes are very different in many respects. The Y chromosome is much smaller than the X chromosome. However, homologous regions are found on both chromosomes, suggesting that they may have evolved from a single homologous pair (*Figure 2.1*). Of these homologous regions, a major region located on the tips of the short arms is the site of an obligate cross-over during male meiosis and is thought to be required for correct meiotic segregation. As DNA sequences within this chromosomal segment do not show strict sex-linked inheritance, it is referred to as the pseudo-autosomal region.

The Y chromosome carries only a few functional genes, including the recently identified *SRY* gene, which maps to Yp in the region adjoining the pseudo-autosomal region [4]. This gene is the major gene involved in determining maleness and is thought to encode the elusive testis-determining factor; rare XX males have been shown to possess the *SRY* gene, presumably as a result of translocation, while XY individuals who lack SRY develop as females. The great bulk of the Y chromosome, however, is genetically inert and is composed of constitutive heterochromatin consisting of different types of satellite DNA (Section 1.8.1). The absence of recombination on most of the Y chromosome has also facilitated studies on human evolution. One recent study of Y chromosome DNA has suggested that most European and Asian men are descended from one of two males [5].

In contrast to the Y chromosome, the human X chromosome contains several thousand functional genes. However, although in human male cells the great bulk of the X chromosome is genetically active, in female cells only one of the two X chromosomes is similarly active. The other replicates late in cell division, remains condensed throughout most of interphase and is cytogenetically visible as a densely staining heterochromatic structure (Barr body or sex chromatin mass). A large proportion of the genes on this chromosome show a comparatively high degree of methylation, consistent with transcriptional inactivity (Section 1.3), and the chromosome is thought to be mostly genetically inert as a consequence of X chromosome inactivation (Lyonization). At an early embryonic stage in female development a random choice is made in each individual cell to inactivate either the paternally inherited or the maternally inherited X chromosome. The same pattern of either paternal or maternal X chromosome inactivation persists in all descendants of the individual cells in which X chromosome inactivation occurred. However, in germ-line cells the originally inactivated X chromosome is reactivated in order that each egg cell has a single active X chromosome.

X chromosome inactivation is thought to have evolved as a form of dosage compensation in cells with more than one X chromosome. Normal cellular processes include interactions between the products of active autosomal and X chromosome genes which depend on their relative dosage. As normal male somatic cells contain only a single X chromosome, the dosage of X genes is necessarily one-half that of autosomal genes. To maintain the same 2:1 dosage ratio between active autosomal and X chromosome genes in female cells, the additional X chromosome requires inactivation. However, a few genes on the inactive X chromosome are known to escape inactivation and are mostly clustered near the pairing region at the tip of Xp, but include at least two genes which map to the X chromosome inactivation centre at Xq13 (*Figure 2.1*). One of these, the *XIST* gene, is only active on inactivated X chromosomes, an observation which is consistent with its suggested role in inducing X chromosome inactivation [6].

2.3 Gene duplication and divergence

In addition to the general gene duplication that arises as a result of whole genome duplication, selective duplication of specific genes can occur by other mechanisms. In the case of interspersed gene families, gene duplication events are thought to arise as a result of DNA transposition through an RNA intermediate (Section 2.7.7). However, such gene copies normally lack functional regulatory sequences present in the original gene and, with very few exceptions, degenerate into pseudogenes. The evolution of many clustered gene families is also thought to have involved the formation of initially identical gene copies by tandem gene duplication (*Figure 2.2*), probably as a consequence of unequal cross-over or unequal sister chromatid exchange (see Section 2.7.2).

If the presence of two identical copies of a specific gene in the haploid genome does not confer any specific selective advantage or disadvantage to reproductive fitness (e.g. through increased or aberrant gene expression as a consequence of increased gene dosage), a duplicated set of genes per haploid genome increases the possibility of acquiring novel function through the effect of divergent mutation. Selection pressure will be exerted on one of the two tandemly duplicated genes to conserve the original coding sequence (and therefore the biological function of the encoded product). As the

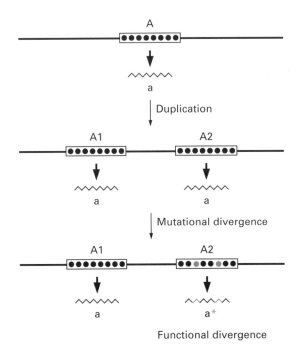

Figure 2.2: *Tandem gene duplication.*

other copy of the gene is effectively redundant, it can acquire mutations in its coding sequence at essentially the same rate as that for the non-coding DNA sequence.

The mutations acquired by the additional gene copy may alter its coding sequence to generate a novel product which may have slightly different biological properties from the original gene product. Alternatively, it may acquire deleterious mutations and become a conventional pseudogene (see Section 1.7.2). For example, tandem duplication of an ancestral DNA segment of approximately 30 kb has been proposed to account for the contemporary organization of the steroid 21-hydroxylase and complement C4 gene cluster (*Figure 2.3*). Both C4 genes are highly polymorphic and function in the serum complement component of the immune system. Sequence divergence has resulted in *C4A* products generally having a greater capacity to process immune complexes but a lower hemolytic activity than *C4B* products. However, of the duplicated 21-hydroxylase genes, only one, *CYP21*, is normally functional; the other gene, *CYP21P*, shows 97% similarity in sequence to *CYP21* but does not appear to express an mRNA product and is therefore classified as a conventional pseudogene. The degeneration of one of the original duplicated 21-hydroxylase genes into a pseudogene could have happened in response to selection pressure to maintain a maximum of one functional 21-hydroxylase gene per haploid genome (i.e. increased gene dosage was selectively disadvantageous). Alternatively, the 21-hydroxylase gene duplication was unsuccessful as a result of chance acquisition of deleterious mutations before any selectively advantageous functional divergence was achieved.

Duplicated genes which express a diverged polypeptide product may also be functionless pseudogenes. Eventually lack of selection pressure to conserve the sequence will result in the acquisition of mutations which will lead to silencing of gene expression. For example, absence of a recognizable function may mean that the

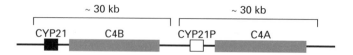

Figure 2.3: *The human 21-hydroxylase/C4 gene cluster.*

expressed θ-globin gene and the chorionic somatomammotropin-like (*CS-L*) gene in the growth hormone gene cluster (see *Figure 1.10*) are examples of expressed pseudogenes. Although the *CS-L* gene shows extensive sequence similarity to the other four genes in the growth hormone gene cluster (>90% matching throughout exons, introns and the immediate flanking sequence), it has a C→A transition in a sequence used by the other four genes as a 5' splice donor site. The mutation results in a different splicing pattern and a novel sequence for the expressed product. If the novel product is advantageous, this gene could be evolving into a functionally different hormone. Presently, the lack of evidence for a novel function suggests that the *CS-L* gene is likely to be a pseudogene.

Some diverged duplicated genes are known to be predominantly expressed in different environments. Sequence divergence in the different genes in the α- and β-globin gene cluster may result in products with slightly different biological properties. The ε-, ζ-, and γ-globin chains could possibly be especially suited to binding O_2 in the comparatively hypoxic environment of early development, whereas the α- and β-globin chains may be the preferred polypeptides in the environment of adult tissues.

Isozymes which are specific for particular subcellular compartments are also often encoded by closely related non-allelic genes on different chromosomes (Section 4.4.3), which are thought to have arisen by ancient gene duplication events. For example, there are two major isoforms of aldehyde dehydrogenase in liver, a cytosolic and a mitochondrial form, which show 68% sequence identity over their 500 amino acid long sequences. The cytosolic and mitochondrial forms are encoded by the *ALDH1* gene on chromosome 9q and the *ALDH2* gene on chromosome 12q respectively. The two genes each have 13 exons and nine out of the 12 introns occur in homologous positions in the two coding sequences, strongly suggesting a common evolutionary origin by some kind of ancient gene duplication event (*Figure 2.4*).

Genes encoding different tissue-specific isozymes also appear to have evolved by a series of gene duplication events. For example, the enzyme alkaline phosphatase is encoded by at least four different genes which show tissue-specific differences in expression. Of these, three are clustered near the telomere of 2q. *ALPI* and *ALPP* encode alternative forms of the enzyme (87% sequence similarity) found in intestine and placenta, respectively, and another gene encodes a placental-like isozyme. Additionally, the *ALPL* gene located near the telomere of 1p encodes an isozyme expressed in liver, bone, kidney and some other tissues and is more distantly related (57% and 52% sequence similarity to the intestinal and placental forms, respectively). The *ALPL* gene which is >50 kb long is more than five times the size of the *ALPI* and *ALPP* genes and has an additional 5' exon. However, the exon/intron organizations are remarkably similar in all three genes and the coding regions of the genes are interrupted

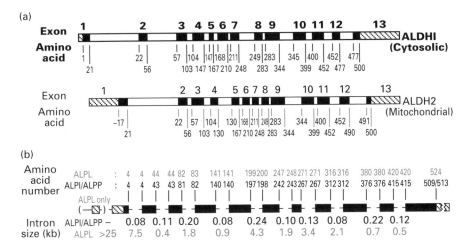

Figure 2.4: *Organization of human genes encoding isozymes of aldehyde dehydrogenase (**a**) and of alkaline phosphatase (**b**).*

by introns at the same positions (*Figure 2.4*). Recent tandem gene duplication events can explain the origin of the 2q genes, whereas the origin of the *ALPL* gene and the precursor to the 2q genes was presumably the result of an ancient genome or gene duplication event followed by subsequent divergence and chromosomal rearrangement.

In general, members of a gene family which are located in the same gene cluster show a higher degree of sequence homology than do members present on different clusters. In the globin gene superfamily, sequence homology between the genes and gene products from different clusters (e.g. α- and β-globin) is much less than between genes and gene products from a single cluster (*Figure 2.5*). Although genes from one cluster have very similar organizations and the intron sizes are largely specific for each cluster, the positions of the two globin introns have been remarkably well conserved; the first intron always interrupts codon 30 or 31, whereas the second intron falls between either codons 99 and 100, 104 and 105, or 105 and 106 (*Figure 2.6*).

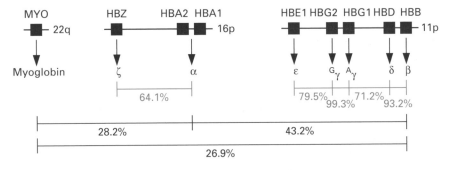

Figure 2.5: *Sequence homology between human globins.*

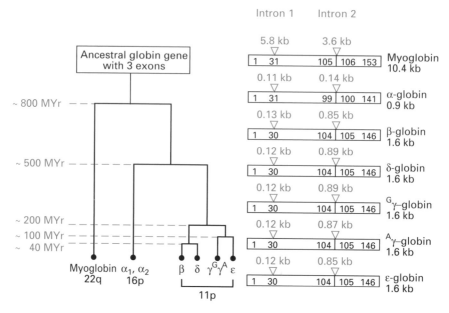

Figure 2.6: *Evolution and exon–intron organization of genes in the human globin gene superfamily. MYr – million years.*

2.4 Exon duplication and exon shuffling

Exon duplication events can create polypeptides with repeating domains which may be functionally advantageous, paticularly for some structural proteins [7]. The sequence of exons in many cases reveals evidence for both evolutionarily recent and ancient intragenic duplications. An illustrative example is provided by the 41 exons of the *COL1A1* gene which encode the part of $\alpha1$(I) collagen that forms a triple helix; each exon essentially encodes an integral number of copies (one to three) of an 18-amino acid motif which is itself composed of six tandem repeats of the structure Gly-X-Y where X and Y are variable amino acids (*Table 2.2*).

In most cases, intragenic duplication events have been followed by substantial nucleotide sequence divergence between the different repeat units, so that the repeated structure may only be revealed by statistical analysis. However, certain genes show very high sequence homology between intragenic repeated coding DNA units (*Table 2.2*). One extraordinary example is provided by genes encoding ubiquitin. This highly conserved 76-amino acid polypeptide is involved in several distinct cellular functions, all of which involve attachment of its C-terminus to free amino groups of other proteins. Of the three types of functional human ubiquitin gene, the coding DNA of the *UbB* and *UbC* genes is not interrupted by introns, and comprises multiple tandem repeats of the 228 bp ubiquitin coding unit (*Figure 2.7*). The initial polypeptide product encoded by the *UbB* and *UbC* genes is a polyubiquitin precursor polypeptide which contains a single unrelated C-terminal amino acid (cysteine or valine). It is presumed to be cleaved post-translationally into ubiquitin monomers. The third class, *UbA* genes, initially encode a fusion protein of ubiquitin and an unrelated tail protein such as the 52-amino acid tail protein of the *UbA52* gene, which is a component of the large ribosomal sub-unit.

Table 2.2: *Examples of intragenic repetitive coding DNA*

Gene(s)	Size of repeat in nucleotides (amino acids)	Number of copies	Nucleotide sequence homology between copies
Ubiquitin (*UbB* and *UbC* genes)	228 (76)	3 (*UbB*) 9 (*UbC*)	High homology
Involucrin	30 (10)	59	High homology for central 39 repeats
Apolipoprotein a	342 (114) = kringle 4-like repeat[a]	37	High homology; 24 of the repeats are identical in sequence
Collagen	54 (18)	57	Low homology but conserved amino acid motifs based on $(Gly-X-Y)_6$
Serum albumin	585 (195)	3	Low homology
Plasminogen	≈ 230 (75–80)	5	Low homology but conserved protein domains (kringles[a])
Proline-rich protein genes	≈ 60 (16–21)	5	Low homology
Tropomyosin alpha chain	126 (42)	7	Low homology
Immunoglobulin ε-chain, C region	324 (108)	4	Low homology

[a]A kringle is a cysteine-rich sequence that contains three internal disulfide bridges and forms a pretzel-like structure.

The separation by introns of defined coding units has the advantage that new combinations of coding units can be brought together by exon shuffling, for example by the occurrence of unequal recombination events within introns (Section 2.7.2). The resulting mosaic genes may then encode polypeptides with novel functions or superior properties which are evolutionarily advantageous. For example, the low density lipoprotein receptor gene (*LDL*) appears to be a mosaic gene with defined sets of exons showing strong sequence homology with different sets of distinct genes (*Figure 2.8*).

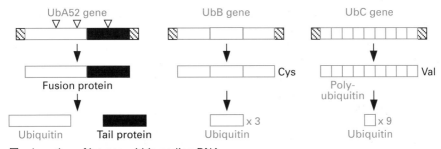

▽ = Location of introns within coding DNA
▨ = Untranslated sequences

Figure 2.7: *Organization and expression of human ubiquitin genes.*

Exon shuffling during protein evolution would not have been expected to alter the reading frame. In accordance with this, type 0 introns (introns which interrupt coding DNA between codons – see *Figure 2.9* for an example) are much more numerous than type 1 or type 2 introns (which interrupt coding DNA after the first or second nucleotide of a codon, respectively), and probably represent the ancestral state.

Implicit in the above rationale for introns is the idea that exons define units of structure or function in proteins. However, exons in contemporary human genes do not

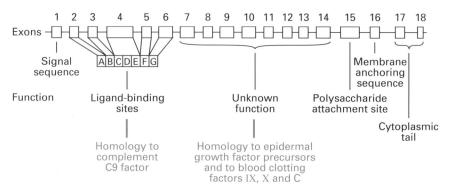

Figure 2.8: *Functional division and intergenic sequence homology of exons in the LDL receptor gene.*

consistently relate to units of structure or function in proteins. In some cases, exons in human genes do not even contain coding DNA, as in the case of the insulin gene where the first intron separates component segments of a single non-coding untranslated region (*Figure 2.9*). Much of the present day relationship of exons to protein structure may nevertheless represent a degenerate state of an ancestral correspondence between exons and structure–function modules in proteins. Any proposed function for introns cannot, however, be a general one because of the small minority of human genes which lack introns, including all mitochondrial genes, most RNA genes and the multiple genes which encode the α- and β-interferons, histones, various hormone receptors and cell receptors, and certain members of the ribonuclease superfamily.

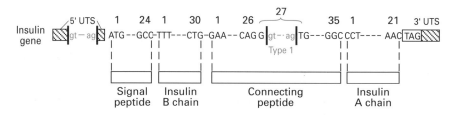

Figure 2.9: *Location of introns in the human β-globin and insulin genes.*

The origin of introns is also uncertain. Although they are common in eukaryotes, they are generally absent in eubacteria but have been found in several genes in archaebacteria. Many human genes have introns which have been considered to be ancient on the basis that their precise intragenic location has been remarkably conserved throughout evolution (*Figure 2.6*). However, the positions of introns in several genes (e.g. those for actin and tubulin) do not appear to have been conserved over long evolutionary time-scales, and may instead have been recently inserted into these genes by DNA transposition.

2.5 Origin of sequence variation and polymorphism

Sequence variation of human gene products reflects alterations of sequences in the nuclear genome and at the level of gene expression, notably RNA processing. At the genome level, differences between allelic sequences at a single chromosomal locus and also differences between related non-allelic sequences at different loci can contribute to sequence diversity. In the former case allelic sequence variation is traditionally described as polymorphism if more than one variant (allele) at a locus occurs in a human population with a frequency greater than 0.01. Although each individual possesses a maximum of only two different alleles at any one locus, a population survey of many individuals may reveal several different alleles at that locus, especially in the case of highly polymorphic loci. The small minority of DNA polymorphisms which lead to amino acid differences contributes to polymorphism at the protein level, as do differential transcription and RNA processing events (see Section 2.8).

In addition to allelic sequence variation, particular types of DNA sequence or protein can show considerable diversity within each individual because of variation in sequence between closely related non-allelic genes which are members of a specific DNA sequence family. For example, both the *C4A* and *C4B* loci are highly poly-morphic and encode the C4 serum complement protein. Allelic sequence variation at each locus produces allotypic variation (e.g. *C4A1*, *C4A2*, *C4A3*, etc. at the *C4A* locus), whereas the presence of two different C4 loci introduces additional isotypic variation (i.e. *C4A* and *C4B* are isotypes). Diversity due to expression of non-allelic genes is most pronounced in the case of the different immunoglobulin and T-cell receptor chains as a result of unique cell-specific DNA rearrangements (see Section 2.7.3). In addition, many examples are known of structural proteins and enzymes which are individually encoded by several different but closely related loci. In the case of such proteins or enzymes, several alternative isoforms (or isozymes) are present in each individual, some of which may be restricted in their expression to specific tissues or subcellular compartments.

Variation in the human genome reflects a variety of chromosomal and genetic mechanisms. Spontaneous aberrations in meiosis, mitosis or fertilization can generate cells with a different number of chromosomes or chromosome types than normal. DNA variation resulting from genetic mechanisms is more subtle. It can arise from a number of processes including mutation induced by environmental agents such as ionizing radiation and chemical mutagens, spontaneous errors in DNA replication or repair, spontaneous or programmed base modification and spontaneous or programmed intragenomic mechanisms of DNA sequence rearrangement and turnover. The latter include a variety of genetic processes which are particularly operative on repeated DNA sequences (see Section 2.7).

While contributing to DNA variation and polymorphism the same mechanisms are also responsible for the phenomenon of concerted evolution, whereby individual members of a specific human repetitive DNA family are often more similar in nucleotide sequence to each other than they are to individual members of the same family in closely related species. Such sequence homogenization is usually most effective for tandemly repeated sequences. For example, sequence homology between two human rDNA genes is generally higher than between a human rDNA gene and a primate rDNA gene. By contrast, the sequence homology between different members of a clustered, interspersed gene family is often less than that between homologs from different mammalian species. Thus the sequence of the human β-globin gene is less related to that of the human ε-globin gene than it is to homologs in chimpanzee, rabbit and mouse (*Table 2.3*).

2.6 Polymorphism due to point mutation

Nucleotide substitutions are the most common type of mutation in coding DNA sequence. Transitions, substitution of a pyrimidine (C or T) by a pyrimidine, or a purine (A or G) by a purine, are more frequent than transversions (substitution of a pyrimidine by a purine or vice versa). The excess of transitions is partly due to the evolutionary instability of methylated cytosine residues. Cytosines which occur in the dinucleotide CpG are often methylated in human and vertebrate DNA to give 5-methylcytosine. Spontaneous deamination of 5-methylcytosine occurs over an evolutionary time-scale to generate thymine [8]. Because the latter is a natural base in DNA, however, it may not be recognized by DNA repair systems as being the product of an aberrant process. The CpG sequence is effectively replaced in this process by TpG (and by CpA on the complementary DNA strand following DNA replication). As a result, the observed frequency of CpG in human genomic DNA is only about 20% of the expected

Table 2.3: *Sequence variation in globin genes*

	Percentage sequence homology			
	Coding DNA	5' + 3' UTS	Introns	Amino acid sequence
Human β-globin/ Chimpanzee β-globin	100	100[a]	98.4	100
Human β-globin/ Rabbit β-globin	89.3	<79[b]	<67[b]	90.4
Human β-globin/ Mouse β-globin[c]	82.1	<66[b]	<61[b]	80.1
Human β-globin/ Human ε-globin	79.1	62	50	75.3

[a]5' UTS only.
[b]Maximum homologies based on counting insertions or deletions of three or more nucleotides as single sites.
[c]Either β^{maj} or β^{min}.

frequency and the CpG dinucleotide appears to be a mutational hotspot and contributes significantly to the molecular pathology of many disorders (Section 5.3).

By comparison, insertion or deletion of even small numbers of nucleotides is comparatively rare in coding sequence DNA; selection pressure is exerted to avoid mutations which alter the translational reading frame of a gene and would result in aberrant gene expression and reduced reproductive fitness. Because of such natural selection pressure to conserve polypeptide sequence and biological function, the frequency of mutation in coding DNA is considerably less than that observed in extragenic or intragenic non-coding DNA. Consequently, the coding DNA component of a specific gene and the derived amino acid sequence show a relatively high degree of evolutionary conservation, as do important regulatory sequences such as the multiple elements of promoters and enhancers. Analogous interspecific comparisons show that untranslated sequences are generally less well conserved than coding DNA, and that introns are the most rapidly diverging gene components (*Table 2.3*). In the latter case, alignment-based comparisons are not facilitated by the relatively high propensity for acquiring insertion or deletion mutations and the estimates of interspecific sequence homology are often crude.

The observed mutations in coding sequence generally also show a pronounced bias towards synonymous ('silent') codon substitutions (i.e. the altered codons specify the same amino acids as the original ones, often by base change at the third base position of codons) or conservative non-synonymous codon substitutions (the altered codons specify new amino acids which, however, are functionally similar to the original ones). On the basis of comparisons between different primate species, the average rate of synonymous codon substitution in the nuclear genome has been calculated to be about $1–2 \times 10^{-9}$ per site per year (i.e. 0.1–0.2% per million years). This is somewhat faster than non-synonymous codon substitution in coding DNA, but is not significantly different from the rate of substitution observed in functionless pseudogenes and extragenic DNA. Silent codon substitutions in human (and other mammalian) genomes are therefore thought to be selectively neutral and they provide a good measure of the natural mutation rate.

The rate and type of substitution varies between different genes and is correlated with the base composition of genes and their flanking DNA. The ubiquitin and histone H4 proteins are the most highly conserved proteins and their genes show a very low rate of non-synonymous codon substitution. In contrast, the rapidly evolving fibrinopeptides and the highly polymorphic HLA antigens show a comparatively high rate of non-synonymous codon substitutions. Although the mean heterozygosity for human genomic DNA has been calculated to be about 0.0037, that is, approximately 1 in 250–300 bases are different between allelic sequences [9], the mean heterozygosity in coding DNA is generally considerably less than that in non-coding DNA. However, the HLA-encoding genes are exceptionally polymorphic and alleles can diverge by as much as 5% of their nucleotide sequence. By comparison, sequence divergence between analogous DNA sequences of humans and closely related primates is generally somewhat less than 2%. For example, comparison of over 4000 bp of DNA sequence flanking the 5' end of the human and chimpanzee β-globin genes revealed only 67 differences, giving an average divergence level of 1.6% [10]. Despite this very close general relationship, certain human genomic DNA sequences are known to be human-specific (i.e. they do not appear to have homologs in any other species) and may contribute to reproductive isolation between man and closely related primates.

Mutation rates have also been observed to vary in different genomic regions, and it has been suggested that mutation rates vary with the timing of replication of different chromosomal regions in the germ line and that such regional differences may possibly explain the origin of isochores (see Section 1.2.1). Additionally, mitochondrial DNA sequences are thought to be evolving at a rate which on average is more than ten times faster than their nuclear DNA counterparts (approximately 2–4% sequence divergence per million years).

2.7 DNA variation due to intragenomic sequence exchange and rearrangements

2.7.1 DNA variation due to gene conversion

The sequences of some alleles at certain gene loci suggest that they originated by replacement of the usual gene sequence, or a segment of the gene sequence, by a similar sized sequence copied from a related non-allelic gene. Such observations suggest the operation in the human genome of a genetic process resembling gene conversion, a non-reciprocal transfer of sequence information between non-allelic or allelic genes. One possible mechanism envisages formation of a heteroduplex between a DNA strand from the donor gene and a complementary strand from the acceptor gene (*Figure 2.10*). Following heteroduplex formation, conversion of an acceptor gene segment may occur by mismatch repair; DNA repair enzymes recognize that the two complementary strands of the heteroduplex are not perfectly matched and 'correct' the DNA sequence of the acceptor strand to make it perfectly complementary in the converted region to the sequence of the donor gene strand. Conversion of a segment within a gene can generate diversity by creating a mosaic gene composed of sequences derived from different alleles or different loci. However, if the converted region encompasses a whole gene, the same mechanism can reduce sequence variation at a locus (polymorphism) and result in homogenization of sequences at different related loci. In mammals and humans, the limited evidence suggests that gene conversions often appear to involve small regions (often a few hundred base pairs or less). Although gene conversion has been well studied in simple eukaryotes such as yeast, formal proof for meiotic gene conversion in humans cannot be obtained (because all the products of individual meioses cannot be recovered) and alternative explanations involving multiple recombination are possible.

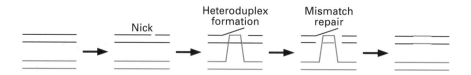

Figure 2.10: *Gene conversion by mismatch repair of a heteroduplex.*

2.7.2 DNA variation due to recombination

Recombination (cross-over) involves breakage of a double-stranded DNA molecule (chromatid) on each of two chromosomes and rejoining of the fragments from the two chromosomes to generate new recombinant strands. Homologous recombination describes recombination occurring at meiosis or mitosis between identical or very similar DNA sequences, normally on non-sister chromatids of homologous chromosomes. Non-homologous recombination (illegitimate recombination) is generally rare and occurs between two sites that show little, if any, homology, and which may be located on two different types of chromosome, or on the same chromosome.

Homologous equal recombination involves cleavage and rejoining of the chromatids at the same position on the chromatids (i.e. recombination occurs between allelic sequences and the breakpoints are in comparable positions in the two alleles). If such allelic recombination occurs within a gene, fusion genes will be formed, consisting of 5' and 3' sequence elements derived from different alleles (*Figure 2.11*). This can generate new alleles and is known to generate polymorphism in the case of certain genes, such as HLA genes. Homologous unequal recombination (unequal cross-over) involves cleavage and rejoining of non-sister chromatids at homologous but non-allelic sequences and always results in deletions and duplications, thereby creating length variation (*Figure 2.11*). Intragenic unequal cross-over introduces additional variation by creating fusion genes. However, unequal cross-over occurring in the interval between clustered genes can result in homogenization of sequences at different loci.

Unequal cross-over can occur between non-allelic members of a regular tandemly repeated DNA family and can generate a variable number of tandem repeats (VNTR) polymorphism. Such VNTR polymorphism is always characterized by alleles which differ in size by integral multiples of the length of the basic repeat unit and has been observed in the case of conventional satellite DNA (where it can give rise to extensive variation in heterochromatin content), minisatellite DNA and microsatellite DNA. Certain tandemly repeated gene families and intragenic tandem repeats also show evidence of VNTR polymorphism and exhibit variation in gene copy number or in gene length, respectively (*Table 2.4*). In the former case, intragenic unequal cross-over results in fusion genes consisting of 5' and 3' sequence elements derived from non-allelic genes. Although VNTR polymorphism is known to result from unequal cross-over in the 21-hydroxylase/C4 gene cluster (Section 5.3.1), VNTR polymorphism at

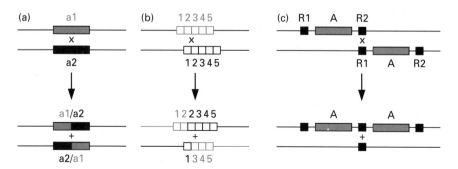

Figure 2.11: *Generation of diversity by recombination between alleles* **(a)**, *non-allelic tandem repeats* **(b)** *and non-allelic interspersed repeats* **(c)**.

other loci may often be attributable to alternative mechanisms (see Sections 2.7.4 and 2.7.5).

Unequal cross-over has also been invoked to explain variation in gene copy number and the origin of fusion genes in many clustered multigene families which do not show the regular tandem repetition of VNTR systems. Additionally, intragenic length variation due to deletion/duplication events resulting from unequal cross-over has been considered to occur in the absence of regular tandem repeats. In these cases unequal cross-over is envisaged to occur between repeated elements which are non-contiguous (*Figure 2.11*). Frequently, for example, the end points of large deletions and duplications have been observed to occur within non-allelic *Alu* sequences which can be located several kilobases apart (Section 5.3.1). Such observations have suggested a general role for *Alu* sequences in promoting recombination. Initial gene duplications in the evolution of clustered multigene families may often have involved an unequal cross-over event between *Alu* repeats or other dispersed repetitive elements [11].

2.7.3 DNA rearrangements in immunoglobulin and T-cell receptor genes

Lymphocyte populations in a single person show a unique degree of diversity in the B-cell immunoglobulins and T-cell receptors. Such diversity is dependent on

Table 2.4: Variation in copy number of human genes and intragenic repeat units

Locus	Nature of repeat	Size of repeat unit	Variation in copy number	Genetic mechanism[a]
FMR-1 (fragile X-linked mental retardation gene)	$(CCG)_n$, i.e. poly-arginine	3 bp	15–65 in normal chromosomes	SR + UEC/UESCE?
Androgen receptor gene	$(CAG)_n$ = poly-glutamine	3 bp	17–26	SR + UEC/UESCE?
Involucrin	Central array of tandemly repeated units of 10 codons	30 bp	37–40	UEC? UESCE?
UbC (ubiquitin)	Poly-ubiquitin coding unit	218 bp	7–9	UEC (UESCE?)
CYP21/C4 gene cluster	Tandemly repeated compound gene cluster	30 kb (CYP21+C4)	1–4	UEC (UESCE?)
rDNA	Tandemly repeated compound gene cluster	43 kb (28S + 5.8S + 18S)	?	UESCE (UEC?)
Amylase gene cluster	Gene cluster	100 kb (3 genes)	1–3 (3–9 genes)	UEC? UESCE?

[a]SR = slippage replication; UEC = unequal cross-over; UESCE = unequal sister chromatid exchange.

programmed rearrangement of the chromosomal DNA in these cells in order to bring particular combinations of coding DNA elements together. In the germ-line, the genes which encode the heavy and light chains of immunoglobulins and T-cell receptors show an unusual organization with clustering of elements which specify similar components (e.g. variable regions) of the mature polypeptide products. As an individual B or T lymphocyte matures, somatic recombination occurs between the different clusters in order to bring into apposition a particular cell-specific but random combination of coding sequence elements.

Different rearrangements may be involved in ensuring the juxtaposition of particular types of coding sequence elements. For example, in the case of κ light chain production, both deletions and also megabase-sized inversions are known to be involved in bringing together different combinations of exons encoding variable regions (V_κ) and exons encoding joining regions (J_κ) and constant regions (C_κ) [12] (*Figure 2.12*). In the case of immunoglobulin heavy chain gene rearrangements, the stage-specific production of different immunoglobulin classes is effected by rearrangements involving different types of constant chain region (class switching). Such class switches are known to occur through an intrachromatid exchange in which non-allelic sequences are brought into opposition by looping out of the intervening DNA [13]. Cleavage of the chromatid at the paired sequences and rejoining brings together the desired elements while the intervening sequence is circularized and subsequently discarded (*Figure 2.13*).

2.7.4 DNA variation due to unequal sister chromatid exchange

Whereas recombination always involves sequence exchange between chromosomes, sister chromatid exchange, which can also occur at meiosis and mitosis, involves breakage and rejoining of sister chromatids within a single chromosome. Homologous equal sister chromatid exchanges occur between identical DNA sequences on the two

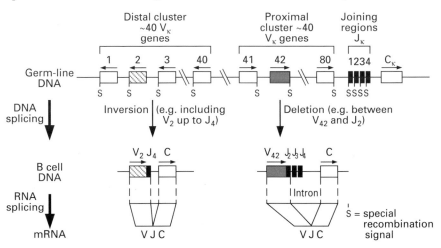

Figure 2.12: *Inversion or deletion results in V–J DNA splicing to produce functional immunoglobulin light chain genes.*

sister chromatids and therefore do not contribute to variation in nucleotide sequence. However, unequal sister chromatid exchange is capable of generating VNTR polymorphism, and variation in copy number of tandemly repeated DNA and clustered repetitive DNA elements (*Figure 2.14*).

In the human ribosomal gene family unequal sister chromatid exchanges appear to be considerably more frequent than unequal cross-over between non-sister chromatids. Additionally, the VNTR polymorphism which characterizes hypervariable minisatellite sequences may be mostly due to within-chromosome unequal sister chromatid exchanges, or possibly in some cases to slippage replication.

2.7.5 DNA variation due to slippage replication

Slippage replication (also called slipped-strand mispairing) can also contribute to VNTR polymorphism by mediating initial expansion and contraction of the size of naturally generated arrays of tandem repeats of simple sequence (less than 10 bp). Unlike unequal sister chromatid exchange and unequal cross-over which necessarily involve mispairing and subsequent sequence exchange between two separate DNA duplexes, slippage replication involves mispairing of the two DNA strands of a single

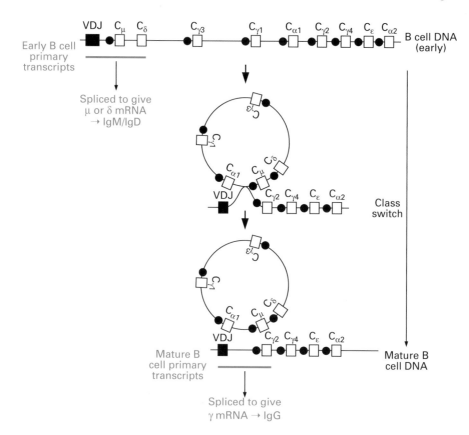

Figure 2.13: *Immunoglobulin heavy chain class switching is mediated by intrachromatid recombination.*

DNA duplex (*Figure 2.14*). However, subsequent expansion and contraction of simple sequence tandem repeats may also occur by unequal cross-over.

In addition to mispairing between tandem repeats, slippage replication has been envisaged to generate large deletions and duplications by mispairing between non-contiguous repeats, and has been suggested to be a major mechanism for DNA sequence evolution [14].

2.7.6 DNA variation due to DNA amplification and saltatory replication

Human tumors often show evidence of DNA amplification events in which cellular oncogenes (see Section 5.4.2) are increased in copy number, sometimes by as much as 1000-fold. In addition, DNA amplification is usually responsible for the high frequency of drug resistance often observed following *in-vitro* selection of many human cell lines exposed to drugs such as methotrexate. The mechanism underlying such DNA amplification is unknown, although one possibility could be multiple successive unequal cross-over events. In normal human cells DNA amplification is rare but *de-novo* amplification of certain human genes has been observed in individuals who have had prolonged exposure to toxic chemicals.

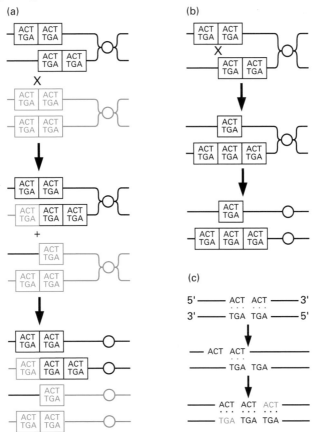

Figure 2.14: *Generation of VNTR polymorphism.* **(a)** *Unequal cross-over.* **(b)** *Unequal sister chromatid exchange.* **(c)** *Slippage replication.* —○— *represents a centromere.*

As well as undergoing length variation as a result of unequal cross-over, satellite DNA sequences in normal cells have been envisaged to be amplified periodically by an uncharacterized process of saltatory replication. The unit that is amplified contains many of the basic tandem repeats which define the satellite DNA. By chance mutation one of the small basic repeat units may contain a diagnostic restriction site, so that amplification of a larger unit containing this repeat can be detected by restriction mapping as a higher order repeat unit (Section 1.8.1).

2.7.7 DNA variation due to DNA transposition

Unlike the extensive genetic variation associated with regions of tandemly repeated DNA in the human genome, polymorphism due to DNA transposition is more limited. One example is known of a human DNA polymorphism which arises through DNA-mediated transfer between autosomes. However, the great majority of DNA transpositions appear to have occurred by retroposition (i.e. RNA-mediated transposition), a process which is thought to account for the origin of processed pseudogenes (*Figure 2.15*). They include active transpositions by a small percentage of the members of the interspersed highly repeated *Alu* and *Kpn (L1)* families, which can contribute to genetic disease (see Section 5.3.5). Insertion polymorphism due to the presence or absence of a presumed transposon has also been observed in a number of other cases. For example, the processed dihydrofolate reductase pseudogene *DHFRP1* is known to be polymorphic, sometimes occurring in human genomic DNA (on chromosome 18) and sometimes being absent.

Gene copies which arise by retroposition lack intron sequences and also important 5' flanking regulatory sequences normally found in the original gene. Consequently, unless they come under the transcriptional control of a sequence at the integration site, they will be transcriptionally inactive. Lack of selection pressure to conserve function will result in rapid sequence divergence. However, a few cases are known of intronless genes which have characteristics of a processed pseudogene but retain functional activity. For example, the intronless *PGK2* gene on 19p encodes a 417-amino acid, testis-specific phosphoglycerate kinase [15]. The *PGK2* product is 87% homologous in sequence to a more widely distributed phosphoglycerate kinase encoded by the X-linked intron-containing *PGK1* gene.

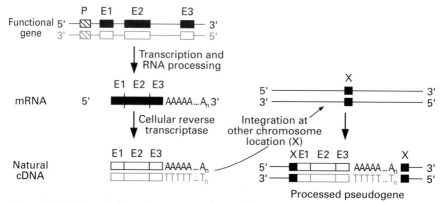

Figure 2.15: *Generation of a processed pseudogene by RNA-mediated DNA transposition. P – promoter; E1, E2, E3 – coding DNA sequences (or copies of such).*

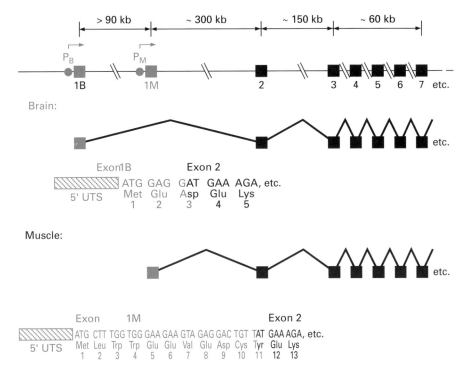

Figure 2.16: *The dystrophin gene is transcribed from different promoters in brain and muscle. P_B, P_M – brain-specific and muscle-specific promoters respectively.*

2.8 Sequence variation due to differential transcription and RNA processing

Many individual human genes show evidence of differently sized RNA transcripts as a result of differential usage of alternative promoter sequences, alternative polyadenylation sites or alternative splicing signals. For example, the human dystrophin gene has two different promoters which are activated in different tissues; a brain-specific promoter activates transcription at a location which is more than 90 kb upstream of the muscle-specific promoter (*Figure 2.16*). Accordingly, the first exon differs in the dystrophin mRNA in brain and muscle resulting in a different N-terminal amino acid sequence. In addition, many individual human genes undergo alternative splicing to yield different mRNA sequences encoding protein isoforms which may be tissue-specific [16]. In some cases, tissue-specific products of divergent function may also derive from a single gene. For example, alternative splicing and polyadenylation of the calcitonin gene results in the synthesis of calcitonin, a circulating Ca^{2+} homeostatic hormone, in the thyroid, and of calcitonin gene-related peptide (CGRP) in the hypothalamus, which may have both neuromodulatory and trophic activities (*Figure 2.17*). Alternative splicing can also regulate protein localization by generating soluble forms of membrane proteins as in the case of many members of the immunoglobulin superfamily, including class I and II HLA genes, IgM and the CD8 gene.

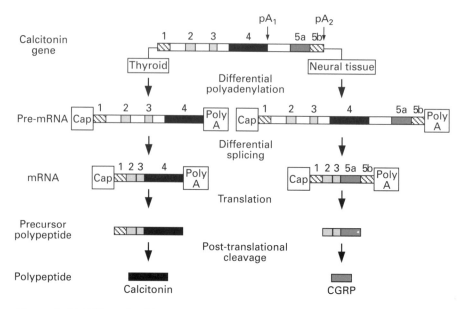

Figure 2.17: *Differential RNA processing results in tissue-specific products of the calcitonin gene. pA1, pA2 – alternative polyadenylation sites. Filled boxes – coding DNA. Open boxes – introns. Hatched boxes – untranslated sequences.*

In addition, a unique form of tissue-specific RNA editing is seen in the case of the apoB lipoprotein gene. In the liver this gene encodes a 14.1 kb mRNA transcript and a 4536-amino acid product, apoB100. However, in the intestine the same gene encodes a 7 kb mRNA which contains a premature stop codon not present in the gene and encodes a product, apoB48, which is identical in sequence to the first 2152 amino acids of apoB100 [17]. The premature stop codon is generated in the intestinal mRNA by a single C→U change at nucleotide 6666.

References

1. Yang, D.Y., Oyaizu, Y., Oyaizu, H., Olsen, G.J. and Woese, C.R. (1985) *Proc. Natl Acad. Sci. USA*, **82**, 4443.
2. Cann, R.L., Stoneking, M. and Wilson, A.C. (1987) *Nature*, **325**, 31.
3. Comings, D.E. (1972) *Nature*, **238**, 455.
4. Sinclair, A.H., Berta, P., Palmer, M.S. *et al.* (1990) *Nature*, **346**, 240.
5. Oakey, R. and Tyler-Smith, C.(1990) *Genomics*, **7**, 325.
6. Brown, C.J., Ballabio, A., Rupert, J.L., Lafreniere, R.G., Grompe, M., Tonlorenzi, R. and Willard, H.F. (1991) *Nature*, **349**, 38.
7. Blake, C.C.F. (1985) *Int. Rev. Cytol.*, **93**, 149.
8. Bird, A.P. (1986) *Nature*, **321**, 209.
9. Cooper, D.N., Smith, B.A., Cooke, H.J., Niemann, S. and Schmidtke, J. (1985) *Hum. Genet.*, **69**, 201.
10. Savatier, P., Trabuchet, G., Faure, C. *et al.* (1985) *J. Mol. Biol.*, **182**, 21.
11. Kudo, S. and Fukuda, M. (1989) *Proc. Natl Acad. Sci. USA*, **86**, 4619.
12. Weichhold, G.M., Klobeck, H.-G., Ohnheiser, R., Combriato, G. and Zachau, H.G. (1990) *Nature*, **347**, 90.
13. von Schwedler, U., Jack, H.-M. and Wabl, M. (1990) *Nature*, **345**, 452.
14. Levinson, G. and Gutman, G.A. (1987) *Mol. Biol. Evol.*, **4**, 203.

15. McCarrey, J.R. and Thomas, K. (1987) *Nature*, **362**, 501.
16. Smith, C.W.J., Patton, J.G. and Nadal-Ginard, B. (1989). *Ann. Rev. Genet.*, **23**, 527.
17. Powell, L.M., Wallis, S.C., Pease, R.J., Edwards, Y.H., Knott, T.J. and Scott, J. (1987) *Cell*, **50**, 831.

Further reading

Darnell, J.E., Lodish, H. and Baltimore, D. (1990) *Molecular Cell Biology,* 2nd edn. Scientific American Books, New York.

Lewin, B. (1990) *Genes IV.* Oxford University Press. Oxford.

Li, W.-H. and Graur, D. (1991) *Fundamentals of Molecular Evolution.* Sinauer Associates Inc., Sunderland, Massachusetts.

3
ANALYZING HUMAN DNA

Because the human genome is so complex, any specific gene or DNA fragment of interest normally represents only a tiny fraction of the total DNA in a cell; for example, the β-globin gene comprises 0.00005% of genomic DNA. In order to study a specific piece of DNA, two major approaches have been applied.

Selective amplification (cloning) of the desired DNA fragment. This can be achieved *in vivo* or *in vitro*. Amplification *in vivo* involves fractionating a complex mixture of DNA fragments by transferring individual fragments into recipient cells (often bacterial or yeast cells), and selectively propagating those cells which contain a desired DNA fragment. This is the conventional approach to DNA cloning. *In-vitro* amplification involves a cell-free method of DNA cloning, the polymerase chain reaction (see Section 3.3).

Specific detection of the desired DNA fragment. This can be achieved if another fragment from the same locus has previously been isolated and purified, for instance by the selective amplification methods described above. The previously isolated fragment, which must share some DNA sequence homology with the fragment to be investigated, can be labeled and used as a probe to identify the desired fragment in a complex mixture of DNA fragments such as total genomic DNA.

3.1 Origin and principle of DNA probes

A DNA probe is any piece of DNA which has been labeled in some way and which is used in a hybridization assay (see Section 3.1.4) to identify other DNA or RNA sequences (target sequences) which are closely related to it in base sequence. DNA probes may be made as single- or double-stranded molecules, but the working probe must be in the form of single strands. Double-stranded DNA probes must therefore be denatured before being used. This is normally achieved by heating the DNA; the weak hydrogen bonds between the complementary strands are disrupted at high temperatures and the strands will dissociate. If the target sequences are initially double-stranded, they too must be denatured before being used in the hybridization assay.

Two major classes of DNA probe are used:

(a) conventional DNA probes, isolated by DNA cloning; these may range from 0.1 to 45 kb in length and are usually (but not always) originally double-stranded; and

(b) oligonucleotide probes, which are short (typically 15–50 nucleotides) single-stranded pieces of DNA synthesized chemically (see Section 3.1.2).

3.1.1 Cell-dependent DNA cloning

Two basic varieties of this method have popularly been undertaken, depending on the nature of the starting DNA.

(a) Genomic DNA cloning. The DNA to be cloned is obtained by cleaving large genomic DNA molecules into pieces of a manageable size, usually from about 0.1 to about 45 kb. This is usually achieved by digesting genomic DNA with a specific restriction endonuclease (see Section 3.2.1).
(b) cDNA cloning. The DNA to be cloned consists of artificially constructed DNA copies of mRNA. Isolated mRNA is used as a template by the enzyme reverse transcriptase initially to synthesize a single-stranded complementary DNA (cDNA) copy of the mRNA, typically 0.1–10 kb long. The single-stranded cDNA is then converted to a double-stranded form for cloning.

The DNA fragments to be cloned are ligated (covalently linked) to specialized vector DNA molecules which are capable of replication in a cell. Cloning vectors are often modified plasmid molecules (small circular double-stranded DNA molecules which are often found in bacterial cells and replicate independently of the host cell chromosome) or modified bacteriophages. The products of the ligation reaction, including hybrid recombinant DNA molecules containing vector plus human DNA, are then used to transform a population of cells. These are usually bacterial cells such as *E.coli*, which are treated so that a fraction of them become competent, meaning that they are capable of taking up foreign DNA from the extracellular environment. Only a small percentage of the cells will take up the human DNA, but those that do usually take up only a single molecule. This is the basis of the critical fractionation step in cell-based DNA cloning; the population of transformed cells can be thought of as a sorting office in which the complex mixture of DNA fragments is sorted by depositing individual DNA molecules into individual recipient cells.

The transformed cells are allowed to multiply and in the case of plasmid-transformed bacterial cells they will form bacterial colonies consisting of clones of identical progeny of a single ancestral cell. At this stage, various screening procedures can be used to select clones carrying recombinant DNA. The resulting collection of cell clones containing recombinant DNA molecules is called a DNA library. For a genomic DNA library, the representation of cloned DNA fragments (DNA clones) does not depend significantly on the tissue of origin of the DNA, but in a cDNA library the range of clones present depends critically on the starting tissue, reflecting the diverse compositions of mRNA populations from different tissues or cell types.

The bacterial colonies can be separated by spreading the population of cells on an agar plate. Selected individual colonies can then be chosen and amplified further in large-scale culture to generate large quantities of a single highly purified human DNA sequence linked to its vector molecule. In most cases, the individual DNA clones contain anonymous DNA, that is, unidentified human DNA. However, DNA libraries can be screened to identify particular DNA sequences if some prior information is available regarding these sequences that will permit their identification.

3.1.2 Oligonucleotide synthesis

Oligonucleotide probes are made by adding mononucleotides, one at a time, to a starting mononucleotide (usually the 3' end nucleotide), which is bound to a solid support. Generally, oligonucleotide probes are designed with a specific sequence chosen in response to prior information about the target DNA. Sometimes, however, oligonucleotide probes are used which are degenerate in sequence; these are made by synthesizing a set of oligonucleotides in parallel which are identical at certain nucleotide positions but differ at others.

3.1.3 Labeling of DNA probes

To be useful as a probe, a DNA sequence must be labeled. Conventional DNA probes are usually labeled by *in-vitro* synthesis. Single strands of the probe DNA are used as templates for synthesis of complementary strands using DNA polymerase and a mixture of the four deoxynucleotides, dATP, dCTP, dGTP and dTTP, at least one of which contains a distinctive labeled group. Oligonucleotide probes are normally labeled by enzymatic addition of labeled nucleotides to one end of the molecule (end-labeling). Traditionally, the label has been a radioisotope (e.g. ^{32}P, ^{35}S or ^{3}H) which can be detected by exposure to X-ray film (autoradiography). Recently, non-isotopic labels have also been used, involving the chemical coupling of a modified reporter molecule, usually biotin, to a nucleotide precursor. The bound probe is then detected using a reporter-binding protein, for example avidin, in a fluorimetric, colorimetric or enzyme assay.

3.1.4 Molecular hybridization

Molecular hybridization assays involve mixing single DNA (or RNA) strands from a labeled probe with those of a test DNA (or RNA) sample, then allowing complementary strands to anneal. Usually the probe comprises one specific small sequence while the target is a heterogenous mixture of many different sequences. The rationale of the method, therefore, is to use the probe to identify any DNA fragments in the test DNA which may be related in sequence to the probe DNA. If the test DNA sample contains any such DNA fragments, heteroduplexes will be formed between probe single strands and complementary single strands of target DNA (*Figure 3.1*). Such heteroduplexes are most stable thermodynamically when the two strands show a high degree of base complementarity (see Section 1.1).

Normally an excess of target DNA over probe DNA is used in order to encourage heteroduplex formation between complementary probe and target sequences. Although heteroduplexes are not as stable as homoduplexes (reannealed probe DNA or reannealed test sample DNA), a considerable degree of mismatching can be tolerated if the overall region of base complementarity is long (>100 bp). Increasing the concentration of NaCl in the hybridization reaction reduces the hybridization stringency and enhances the stability of mismatched heteroduplexes. Conversely, increasing the temperature at which the single strands are allowed to reassociate increases the hybridization stringency by encouraging dissociation (denaturation, or 'melting') of mismatched heteroduplexes. If the region of base complementarity is small, as with oligonucleotide

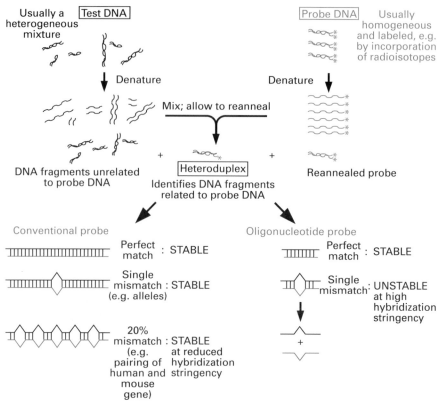

Figure 3.1: *DNA hybridization assay.*

probes (typically 15–20 nucleotides), hybridization conditions can be chosen such that a single mismatch renders a heteroduplex unstable (*Figure 3.1*).

3.1.5 Southern blot hybridization

In this procedure, the test DNA is digested with one or more restriction nucleases, size-fractionated by agarose gel electrophoresis, denatured, and transferred to a nitrocellulose or nylon membrane for hybridization (*Figure 3.2*). During the electrophoresis, DNA fragments, which are negatively charged because of the phosphate groups, are repelled from the cathode (negative electrode) towards the anode (positive electrode), and sieve through the porous gel. Smaller DNA fragments move faster. For fragments between 0.1 and 20 kb long, the migration speed depends on fragment length and scarcely at all on the base composition: fragments in this size range are therefore fractionated by size. Following electrophoresis, the test DNA fragments are denatured in strong alkali.

As agarose gels are fragile, and the DNA in them can diffuse within the gel, it is usual to transfer the denatured DNA fragments by blotting onto a durable nitrocellulose or nylon membrane, to which single-stranded DNA binds readily. The individual DNA fragments become immobilized on the membrane at positions which are a

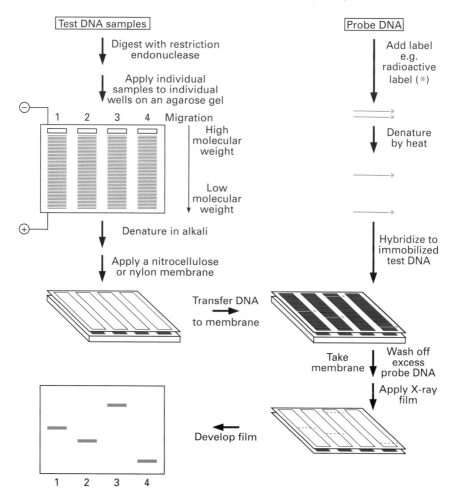

Figure 3.2: Southern blot hybridization.

faithful record of the size separation achieved by agarose gel electrophoresis. Subsequently, the immobilized single-stranded target DNA sequences are allowed to associate with labeled single-stranded probe DNA. The probe will bind only to related DNA sequences in the target DNA, and their position on the membrane can be related back to the original gel in order to estimate their size. Northern blot hybridization is a variant in which the target is RNA instead of DNA.

3.1.6 Dot-blot hybridization

In this procedure the test DNA or RNA is immobilized on a membrane without prior size fractionation. An aqueous solution of the denatured test DNA, for example, total genomic DNA, is simply spotted onto a nitrocellulose or nylon membrane in order to permit binding of DNA to the membrane.

3.2 Using probes to study small DNA segments

Conventional DNA probes have been used in hybridization studies to obtain short-range restriction mapping information concerning a specific locus and to detect a variety of different types of polymorphism.

3.2.1 Restriction mapping

This general procedure involves cutting a piece of DNA with one or more of a series of different bacterial restriction endonucleases and separating the resulting fragments according to size by agarose gel electrophoresis. Type II restriction endonucleases are used. These enzymes recognize specific short sequence elements in double-stranded DNA and then cleave the DNA within, or very close to, the recognition site. A recognition site for such enzymes, a so-called restriction site, is a short stretch of double-stranded DNA sequence, often between 4 and 8 bp in length, which is usually palindromic, meaning that the sequence of bases within the recognition site is the same on both DNA strands when read in the 5'→3' direction. Because of the overall deficiency of G + C in the human genome and the conspicuous deficiency in the CpG dinucleotide, recognition sites that are GC-rich will occur comparatively less frequently than expected (*Table 3.1*).

Double digests (cleavage by two different enzymes) and partial digests (reduced digestion so that not every cleavage site is actually cut) help in relating the different restriction fragments to each other. The resulting information can be used to construct

Table 3.1: Restriction nucleases

Enzyme	Source	Sequence cut	Average expected fragment size (kb) in human DNA[a]
*Alu*I	*Arthrobacter luteus*	AGCT	0.3
*Hae*III	*Hemophilus aegyptus*	GGCC	0.6
*Taq*I	*Thermus aquaticus*	T<u>CG</u>A	1.4
*Hpa*I	*Hemophilus parainfluenzae*	C<u>CG</u>G	3.1
*Mnl*I	*Moraxella nonliquefaciens*	CCTC/GAGG	
*Hind*III	*Hemophilus influenzae* Rd	AAGCTT	3.1
*Eco*RI	*Escherichia coli* R factor	GAATTC	3.1
*Bam*HI	*Bacillus amyloliquefaciens* H	GGATCC	7
*Pst*I	*Providencia stuartii*	CTGCAG	7
*Mst*I	*Microcoleus* species	CCTNAGG[c]	7
*Sma*I	*Serratia marescens*	CC<u>CG</u>GG	78
*Eag*I	*Enterobacter agglomerans*	<u>CG</u>GC<u>CG</u>	390[b]
*Sac*I	*Streptomyces achromogenes*	CC<u>GCGC</u>	390[b]
*Bss*HII	*Bacillus stearothermophilus*	G<u>CGCGC</u>	390[b]
*Not*I	*Norcadia otitidis-caviarum*	G<u>CG</u>GC<u>CG</u>C	9766[b]

[a]Assuming 40% G+C, and a CpG frequency 20% of that expected.
[b]Observed average sizes appear to be lower: about 100–200 kb for *Eag*I, *Sac*I, *Bss*HII and about 1000–1500 kb for *Not*I.
[c]N = A, C, G or T.

a restriction map, a linear map of the relative positions of recognition sites for a variety of restriction nucleases. If the restriction maps of two independently isolated human DNA fragments show extensive sharing of restriction sites, it is highly likely that the two fragments contain overlapping DNA sequences or are closely related members of a repeated DNA sequence family.

3.2.2 Direct detection of pathological point mutations by restriction mapping

Once a specific gene has been cloned, it can be used as a probe to identify abnormalities in that gene in the DNA from any human source. Occasionally, a pathogenic point mutation coincidentally destroys a recognition site (or creates a new recognition site) for a specific restriction endonuclease. In such cases, it is possible to distinguish the normal gene allele from the disease allele by digesting individual genomic DNA samples with the relevant restriction endonuclease and using a gene probe to identify the different-sized fragments characteristic of the normal and disease alleles. For example, sickle cell disease is due to a single nucleotide substitution (A→T) at codon 6 in the β-globin gene; the codon GAG which specifies glutamic acid is replaced by the codon GTG which specifies valine. The sickle cell mutation coincidentally abolishes an *Mst*II restriction site (CCTGAGG) which spans codons 5–7; this sequence is changed to CCTGTGG which is not recognized by *Mst*II (general recognition sequence = CCTNAGG, where N = A, C, G or T). However, the nearest flanking *Mst*II restriction sites (a CCTTAGG site 1.2 kb upstream in the 5' flanking region and a CCTTAGG site 0.2 kb downstream at the 3' end of the first intron) are well-conserved.

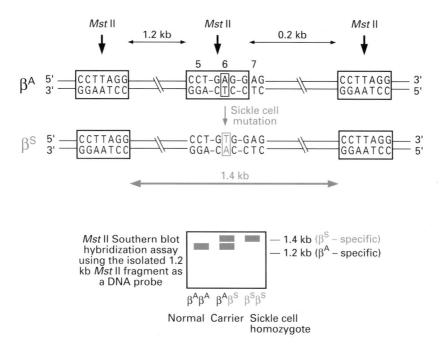

Figure 3.3: *The sickle cell mutation destroys an* Mst*II site and generates a disease-specific restriction fragment.*

Consequently, digestion of genomic DNA with *Mst*II will permit distinction of the normal adult β^A-globin allele (1.2 kb + 0.2 kb *Mst*II fragments) from the sickle cell β^S-globin allele (1.4 kb *Mst*II fragment). By hybridizing a β-globin DNA probe to a Southern blot of *Mst*II-digested genomic DNA it is possible to assay the sickle cell mutation directly and to distinguish normal homozygotes ($\beta^A\beta^A$), carriers ($\beta^A\beta^S$), and sickle cell patients ($\beta^S\beta^S$) (*Figure 3.3*).

3.2.3 Detection of conventional RFLPs

As detailed in Chapter 2, the great majority of mutations are not associated with disease, but are often neutral mutations within non-coding DNA sequences. As a large number of recognition sequences are known for Type II restriction nucleases, many point mutation polymorphisms (see Section 2.6) will be characterized by alleles which possess or lack a recognition site for a specific restriction nuclease and therefore display restriction site polymorphism (RSP). RSPs normally have two detectable alleles (one lacking and one possessing the specific restriction site). In a manner analogous to the preceding section, RSPs can be assayed by digesting genomic DNA samples with the relevant restriction endonuclease and identifying specific restriction fragments whose lengths are characteristic of the two alleles, so-called restriction fragment length polymorphisms (RFLPs) (*Figure 3.4*).

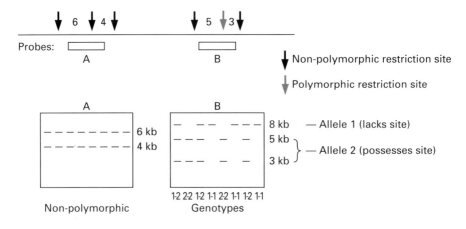

Figure 3.4: *Assay of conventional (RSP-based) RFLPs.*

3.2.4 Detection of point mutations by allele-specific oligonucleotide probes

A point mutation which does not produce a restriction site change may still be detected using allele-specific oligonucleotide (ASO) probes. ASO probes are typically 15–20 nucleotides long and are normally employed under hybridization conditions at which the DNA duplex between probe and target is only stable if there is perfect base complementarity between them. A single mismatch between probe and target sequence is sufficient to render the short heteroduplex unstable (*Figure 3.1*). Oligonucleotide probes can therefore be designed to hybridize to specific alleles of a gene which differ by a single nucleotide at a diagnostic site (*Figure 3.5*). Although ASOs can be used in conventional Southern blot hybridization, it is more convenient to use them in dot-blot assays.

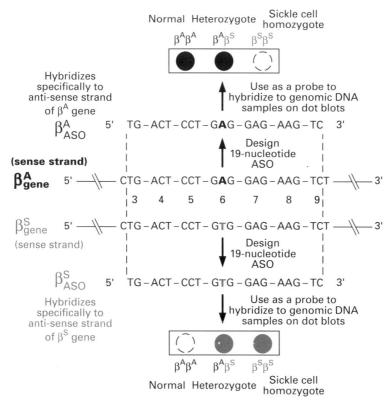

Figure 3.5: *ASO-dot blot hybridization to detect the sickle cell mutation.*

3.2.5 VNTR-based RFLPs and DNA fingerprinting

DNA probes can also be used to monitor VNTR polymorphism (see Section 2.7.2) in human DNA by detecting VNTR-based RFLPs. Genomic DNA samples are digested with a restriction nuclease which recognizes well-conserved restriction sites flanking a VNTR locus, thereby liberating allelic fragments whose lengths differ by integral numbers of repeat units. The resulting VNTR-based RFLPs can be identified by Southern blot hybridization using a suitably locus-specific probe, often derived from a DNA sequence flanking the VNTR repeats (*Figure 3.6*).

If the VNTR locus is a member of a repeated DNA family, the use of a VNTR repeat probe, rather than a unique flanking probe will produce a complex polymorphic pattern. For example, hypervariable minisatellite DNA clones (see Section 1.8.2) have been used as probes against Southern blots of appropriately digested genomic DNA. Cross-hybridization of such probes with the other members of this highly repeated DNA family results in a pattern of hybridization bands representing the summed contributions of two alleles at each of many hypervariable loci scattered throughout the genome. Consequently, the overall polymorphism of the multilocus hybridization

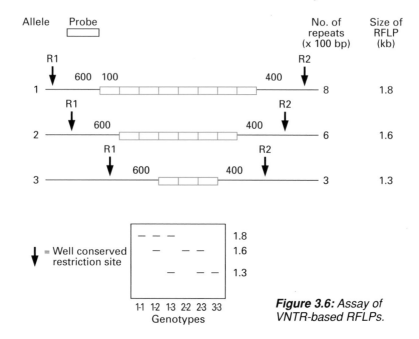

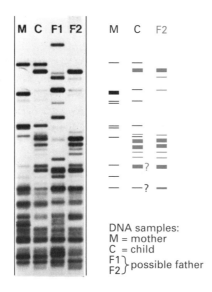

Figure 3.6: *Assay of VNTR-based RFLPs.*

patterns is uniquely high [1]. Because it permits distinction between any two individuals who are not identical twins, probing with hypervariable minisatellites has been termed 'DNA fingerprinting' and has been used widely in forensic science and in settling paternity disputes. The latter application is illustrated in *Figure 3.7*. Hybridizing bands which are found in the child, but not in the mother are not always present in the F_1 sample but are consistently found in the F_2 sample, indicating that F_2 is the father.

DNA samples:
M = mother
C = child
F1
F2 } possible father

Figure 3.7: *DNA fingerprinting in paternity testing. M, mother; C, child; F1 and F2, possible fathers. Photograph courtesy of Cellmark Diagnostics, Abingdon, Oxfordshire, UK.*

3.2.6 Detection of gene deletions by restriction mapping

Certain diseases are associated with a high frequency of deletion of a gene or of part of a gene (see Section 5.3.1). If a partial restriction map has been established for the gene under investigation, deletions can be screened by Southern blot hybridization using an appropriate intragenic DNA probe. If the deletion is a small one, for example, a few hundred base pairs, it is often apparent as a consistent reduction in size of normal restriction fragments in the gene. An individual who is homozygous for this mutation, or is a heterozygote with one normal allele and another with a small deletion, can easily be identified by detecting the aberrant size restriction fragments.

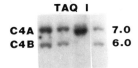

F = Father
M = Mother
P = 21-hydroxylase deficiency patient
N = Normal homozygote

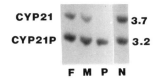

Figure 3.8: *Detection of heterozygous and homozygous gene deletions associated with 21-hydroxylase deficiency. F, father; M, mother; P, 21-hydroxylase deficient patient; N, normal homozygote.*

Large deletions will lead to absence of specific restriction fragments. Homozygous deletion of large DNA segments can easily be detected as complete absence of appropriate restriction fragments associated with the gene. If, however, an individual is heterozygous for a relatively large gene deletion, the deletion may still be detected by demonstrating comparatively reduced intensity of specific gene fragments. For example, patients with 21-hydroxylase deficiency often have deletions of about 30 kb of the 21-hydroxylase/C4 gene cluster. Such pathological deletions eliminate the functional 21-hydroxylase gene, *CYP21*, and an adjacent *C4B* gene, leaving the related *CYP21P* pseudogene and a *C4A* gene. Patients with homozyogous deletions will show absence of diagnostic restriction fragments associated with *CYP21* and *C4B* while carriers of the deletion will show a 2:1 ratio of *CYP21P:CYP21* and of *C4A:C4B* (*Figure 3.8*).

3.3 Using the polymerase chain reaction to study small DNA segments

3.3.1 Basis of the polymerase chain reaction

Polymerase chain reaction (PCR) is a rapid and versatile cell-free method for selectively amplifying defined target DNA sequences present within a heterogeneous collection of DNA sequences (e.g. total genomic DNA) [2,3]. In order to amplify a specific target sequence using PCR, some prior DNA sequence information about the target DNA locus is normally required. This information is necessary to construct two

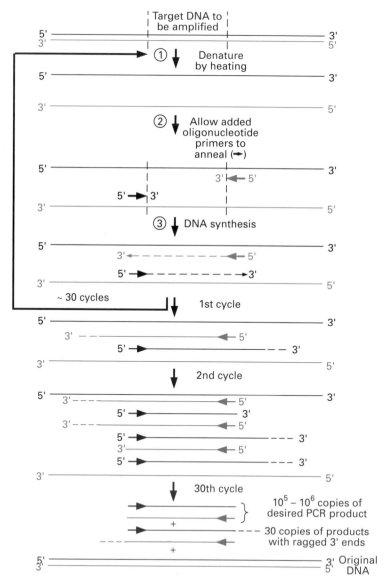

Figure 3.9: *The polymerase chain reaction.*

amplimers, oligonucleotide primer sequences (often 20–30 nucleotides long) which when added to denatured genomic DNA will specifically bind to complementary DNA sequences immediately flanking the desired target region (*Figure 3.9*). The amplimers are designed so that, in the presence of a suitably heat-stable DNA polymerase (often Taq polymerase) and DNA precursors (the four deoxynucleotide triphosphates, dATP, dCTP, dGTP and dTTP) they can initiate the synthesis of new DNA strands which are complementary to the individual DNA strands of the target DNA segment, and which will overlap each other.

PCR is a chain reaction because newly synthesized DNA strands will act as templates for further DNA synthesis in subsequent cycles. After about 30 cycles of successive steps of denaturation, annealing of amplimers and DNA synthesis, the products of the PCR will include, in addition to the starting DNA, about 10^5–10^6 copies of the specific target sequence, an amount which is easily visualized as a discrete band of a specific size on agarose gel electrophoresis. One major limitation of PCR as a DNA cloning method, however, is that it becomes progressively more difficult to amplify large DNA sequences and the method has largely been confined to amplifying small DNA regions (generally less than 5 kb).

PCR is a robust technique which can permit rapid amplification of DNA sequences, even when the starting material contains badly degraded DNA. As a result it has recently been employed in molecular anthropology studies, for example, the analysis of DNA recovered from mummified individuals, and has been used successfully to amplify DNA from formalin-fixed tissue samples. It is also extraordinarily sensitive and permits amplification and study of single-copy DNA starting from a very modest amount of material, even a single cell. The latter property has found numerous applications in forensic science, in diagnosis (Section 6.3), in genetic linkage analysis using single-sperm typing (Section 4.1.5), and suchlike.

In molecular pathology studies the sensitivity of PCR has also been of great advantage. For example, large genes such as the dystrophin gene cannot easily be analyzed for point mutations at the DNA level while dystrophin mRNA, although much smaller, is preferentially expressed in relatively inaccessible tissues such as muscle and brain. However, because of the phenomenon of illegitimate transcription (whereby at least a few mRNA molecules are transcribed from genes in all tissues [4]), blood cells will contain a very small amount of dystrophin mRNA. Blood cell mRNA can be converted to cDNA and dystrophin cDNA sequences selectively amplified by PCR.

3.3.2 Use of PCR to study RSP and VNTR polymorphism

PCR has been used as an alternative to Southern blot hybridization to detect mutations which introduce RSPs or VNTR polymorphism. In both cases DNA sequence information is required to construct oligonucleotide primers which flank the location of the polymorphic site and can direct the synthesis of amplified DNA spanning it. To detect RSPs, the amplified DNA is digested with the relevant restriction nuclease and the products are fractionated by agarose gel electrophoresis in order to detect the presence or absence of the restriction site in the amplified DNA.

In the case of VNTR polymorphism, the use of flanking amplimers can permit PCR amplification of alleles whose sizes differ by integral repeat units (*Figure 3.10*). If the

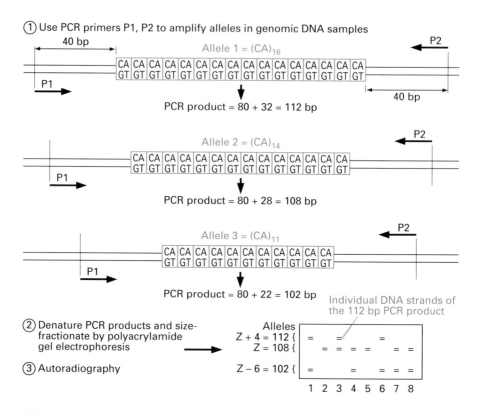

Figure 3.10: *Typing (CA)$_n$ repeats. The most frequent allele length is commonly named 'z' and other alleles are referenced against this.*

VNTR locus is not large, as is the case with individual $(CA)_n/(TG)_n$ microsatellite loci, the PCR products can then simply be size-fractionated by polyacrylamide gel electrophoresis. The PCR normally includes a radioactive nucleotide precursor which becomes incorporated into the products and facilitates their detection by autoradiography. To ensure adequate size fractionation of alleles, the PCR products are denatured prior to electrophoresis. However, because of the peculiarly skewed base composition of the two strands of individual alleles, two bands are normally seen per allele, representing the different mobilities of the two complementary strands.

3.3.3 Other uses of PCR for studying small DNA segments

PCR has also been used to permit rapid DNA sequencing procedures (Section 3.4). Additionally, it has been employed to detect single base changes in DNA sequences. For example, a single base change can be screened by the amplification refractory mutation system (ARMS). In this system a PCR is designed in which one primer hybridizes to the genomic DNA at the site of the possible base change, so that its 3' end nucleotide must pair with the specific base that is to be screened [5]. The PCR will proceed only if there is no base mismatch at the 3' end of the primers (see *Figure 6.8* for an example).

3.4 DNA sequencing

DNA sequencing involves determination of the linear order of the bases of DNA by chemical or, more popularly, by enzymatic methods, notably those based on the use of dideoxynucleotides. These are nucleotide analogs which act as inhibitors of chain extension during DNA synthesis. In the latter approach the usual first step is to prepare a single-stranded DNA form of the DNA to be sequenced (often by cloning the desired DNA region into specialized vector molecules which can be induced to produce single-stranded DNA). A chemically synthesized oligonucleotide primer is annealed to the single-stranded DNA template in the region immediately flanking the segment which is to be sequenced (*Figure 3.11*). The primer then serves to initiate the synthesis of new DNA strands by DNA polymerase in four parallel reactions.

Each reaction contains all four nucleotides, dATP, dCTP, dGTP and dTTP, one of which carries a radioactive isotope, together with one of the four dideoxy analogs, ddATP, ddCTP, ddGTP and ddTTP. Consequently, a specific ddNTP chain terminator competes with its corresponding dNTP for inclusion in the growing DNA chain. As a result, the products of each reaction are a set of DNA chains of different lengths each terminating in a ddNTP. For example, in *Figure 3.11* the C reaction products are 22, 25, and 31 nucleotides long (20 nucleotide primer plus 2, 5 or 11 nucleotides before a ddC residue becomes incorporated). Individual reactions are loaded in neighboring wells of a denaturing polyacrylamide gel and size-fractionated by electrophoresis. On conventional sequencing gels about 200–400 nucleotides can be determined from a single DNA sample.

Recently, a variety of methods have been developed to enable DNA sequences to be obtained directly and conveniently from PCR amplification products. The comparative rapidity of PCR sequencing approaches makes them ideally suited to the direct detection of DNA polymorphism.

3.5 Other methods for detecting single base changes

The methods used to detect single base changes depend on whether prior information is available concerning the identity of the mutation. If no such prior information is available, it is still possible to screen for base changes by direct DNA sequencing of the region of interest. Depending on the size of the DNA region under investigation this approach can be laborious and time-consuming. Accordingly, several scanning methods have been developed to detect single base changes and identify their approximate locations [6]. Subsequently, the exact location of the base change can be identified by DNA sequencing.

3.5.1 Ribonuclease A cleavage

This method uses an RNA probe which hybridizes specifically to the DNA region of interest. Such a probe can be generated by subcloning a conventionally cloned human DNA fragment into a specialized vector. The labeled RNA probe is then annealed to a test sample of either genomic DNA or mRNA, and the enzyme ribonuclease A (RNase A) is added to the reaction. If there is a base mismatch between the RNA probe and its complementary target sequence, the RNA strand will appear to be single-stranded at this position and will be susceptible to cleavage by RNase A.

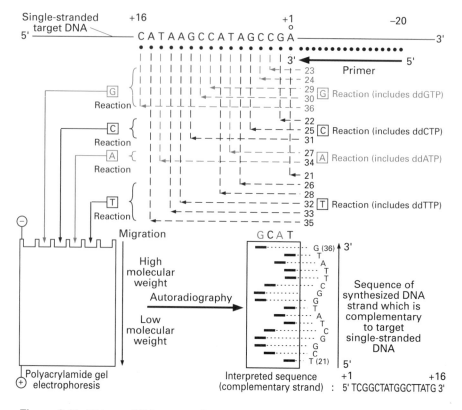

Figure 3.11: *Dideoxy DNA sequencing.*

3.5.2 Denaturing gradient gel electrophoresis

In this method (abbreviated to DGGE), DNA duplexes are forced to migrate through an electrophoretic gel in which there is a gradient of increasing amounts of a denaturant. Migration continues until the DNA duplexes reach a position in the gel where the strands melt and separate, after which the denatured DNA does not migrate any further. A single base pair difference between a normal and a mutant DNA duplex is sufficient to cause them to migrate to different positions in the gel.

3.5.3 Chemical cleavage of mismatch

This method is based on the observation that certain chemicals can react with single-stranded bases of DNA to make them susceptible to cleavage by piperidine. The test DNA and a labeled double-stranded DNA probe for the region of interest are denatured and allowed to reanneal to form heteroduplexes containing a labeled probe DNA strand and a complementary target DNA strand. If there is a single base mismatch between the probe and target strands, detection is possible using hydroxylamine (which preferentially reacts with mismatched C bases) or osmium tetroxide (which reacts with mismatched T bases). In the case of mismatched A or G bases, a probe of the opposite sense (i.e. a complementary strand probe) is used to permit their detection on the opposite strand as mismatched T or C bases, respectively. After chemical treatment, the heteroduplexes are reacted with piperidine and then electrophoresed on denaturing polyacrylamide gels.

3.5.4 Single strand conformation polymorphism

Single-stranded DNA has a tendency to fold up and form complex structures stablilized by weak intramolecular bonds, notably base-pairing hydrogen bonds. The electrophoretic mobilities of such structures on non-denaturing gels will depend not only on their chain lengths but also on their conformations, which is dictated by the DNA sequence. Single strand conformation polymorphism (SSCP) analysis is most conveniently performed by PCR amplification of a desired region of genomic DNA to produce a labeled product (either by using end-labeled PCR primers, or by incorporation of labeled nucleotides during the PCR reaction). Amplified DNA samples are denatured and loaded on a non-denaturing polyacrylamide gel. A test DNA sample that differs by a single base from a standard DNA sample can be identified by a comparative mobility shift during electrophoresis, if the base change results in a change of conformation.

3.6 Studying DNA at the megabase level

3.6.1 Isolation of large DNA fragments

In order to study large human genes and gene clusters, or to map large sections of DNA, it is desirable to be able to clone large DNA fragments. Modified plasmid and bacteriophage lambda vectors, although popularly used for general DNA cloning, are restricted by their limited capacity for accepting large foreign DNA fragments. Instead, large DNA fragments have been cloned mainly using cosmid vectors or yeast artificial chromosomes (YACs).

Cosmid vectors are specialized plasmid vectors which contain the short *cos* sequences of bacteriophage lambda. The latter impose a size limit of 37–52 kb for DNA which can be packaged within bacteriophage lambda protein coats. *In-vitro* packaging of cosmid recombinants into lambda protein coats is an efficient method of transferring foreign DNA into *E. coli* cells and this cloning procedure can normally accommodate DNA fragments from 30 to 46 kb, a size range which will accommodate the majority of human genes. Nevertheless, cosmid vectors have two main disadvantages for cloning human DNA fragments. First, the maximum size of DNA fragments they can accept is still too small to accept some large human genes or gene clusters. Secondly, they require propagation in bacterial cell hosts, and certain types of human DNA fragments contain sequence elements (direct repeats, etc.) which render them unstable in prokaryotic hosts such as bacterial cells.

In order to overcome the limitations of cosmid vectors, a system has recently been devised for cloning large eukaryotic DNA fragments in a simple eukaryotic host, yeast. YAC vectors are plasmid vectors into which have been inserted short DNA sequences which perform essential chromosomal functions in yeast cells [7]. These include centromeres, telomeres and an autonomously replicating sequence (ARS) element which permits independent DNA replication. Suitably cleaved YAC vector molecules can be ligated *in vitro* to large restriction fragments of human DNA (see Section 3.6.2) to generate an artificial yeast chromosome which contains up to 1 Mb of human DNA.

3.6.2 Long range restriction mapping

The organization of DNA sequences spanning up to a megabase or more can be studied by this method. In order to obtain suitably large DNA fragments, DNA is isolated in such a way as to minimize artificial breakage of the large molecules, and the high molecular weight DNA is cleaved by digesting with specialized 'rare-cutter' restriction nucleases. To prepare high molecular weight DNA, samples of cells, for example, white blood cells, are mixed with molten agarose and the resulting mixes are transferred into moulds in which the human cells become entrapped in solid agarose blocks (*Figure 3.12*). The agarose blocks are incubated with hydrolytic enzymes which diffuse through the small pores in the agarose and digest cellular components, leaving the high molecular weight chromosomal DNA virtually intact. Individual blocks containing purified high molecular weight DNA can then be incubated in a buffer containing a rare-cutter restriction nuclease, usually an enzyme with a 6 or 8 bp recognition site containing one or two of the rare CpG dinucleotides. The restriction fragments produced typically average several hundred kilobases in length (*Table 3.1*).

Such restriction fragments are too large to be separated by conventional agarose gel electrophoresis but can be size-fractionated by pulsed field gel electrophoresis (PFGE). This procedure separates large DNA molecules by intermittently forcing them to change their conformation and direction of migration during their passage through an agarose gel [8]. Agarose blocks containing the large DNA fragments are

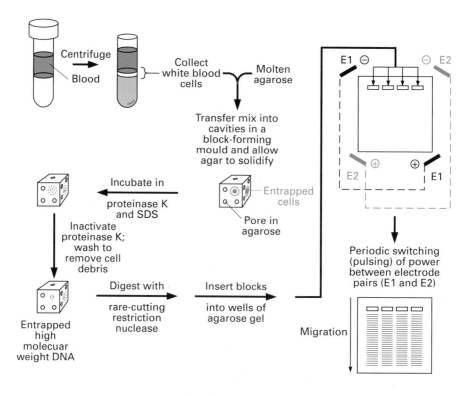

Figure 3.12: *Fractionation of high molecular weight DNA from blood cells by pulsed field gel electrophoresis.*

placed in wells at one end of an agarose gel and the DNA migrates in the electric field. However, during electrophoresis the relative orientation of the gel and the field is altered periodically, typically by setting a switch to deliver brief pulses of power, alternatively activating two differently oriented fields (*Figure 3.12*). Variant techniques use a single electric field but with periodic reversals of the polarity (field inversion gel electrophoresis), or periodic rotation of the gel or electrodes.

PFGE has been used to construct long range restriction maps of large human DNA fragments cloned in YAC vectors. Uncloned genomic DNA can also be mapped by digesting high molecular weight genomic DNA with a panel of rare-cutting restriction nucleases (usually including double digests), and size-fractionating on PFGE gels. The apparently continuous smear of restriction fragments in the gel can be transferred to a nitrocellulose or nylon membrane by Southern blotting for hybridization to a labeled DNA probe. Given a small labeled human DNA probe, therefore, it is possible to generate a restriction map extending over a megabase region at the probe locus.

3.7 Studying gene expression and gene function

The expression of isolated human genes can be studied following transfection into cultured cell lines or, following the incorporation of the gene into the germ-line of an animal, in whole organisms. Such systems permit the expression of normal genes, mutant genes isolated from patients with genetic abnormalities, and also genes which have been altered artificially prior to transfection into the appropriate expression system. Often the isolated human gene is modified *in vitro* by site-directed mutagenesis, by which specific sites in the gene, even single nucleotides, can be altered in a predetermined way to investigate the effect on gene expression. In this way it is possible to map specific gene functions to specific gene domains and to explore the possibility of designing novel genes (designer genes) whose products may have superior biological properties when compared to the natural human genes.

Various expression systems can be used to study the expression of a human gene. Often the gene is cloned into an expression vector and transfected into cultured cells where it may be expressed. The recipient cells may be derived from a different species, often mouse, so that the human gene product can be recognized and studied against the background of the host cell products. The size of the human gene often dictates which expression system can be used. Small genes can be cloned into plasmid or viral expression vectors, while large genes or gene clusters can be inserted into YAC vectors (Section 3.6.1) before incorporation into recipient cells. Very large genes or gene clusters can be incorporated into foreign cells by chromosome-mediated gene transfer [9], a technique in which whole human chromosomes are transferred into a foreign cell by cell fusion, whereupon subchromosomal amounts of the human DNA may be incorporated into the genome of the alien cell.

Certain vectors, notably modified retroviruses, facilitate incorporation of a transfected gene into the chromosomal DNA of the recipient cell (integration) and may lead to stable expression of the integrated gene(s). Non-integrated transfected genes (e.g. genes inserted in vectors which replicate extrachromosomally) are expressed only transiently. Transient expression systems are often used for assaying transcriptional functions such as the strengths of promoters or enhancers.

Transfer of human DNA into the germ-line of an animal permits the creation of a transgenic animal in which the human DNA is stably incorporated into the chromosomal DNA of the animal. Expression of a human gene in a transgenic animal allows many aspects of gene expression and gene regulation to be investigated. Additionally, incorporation of foreign DNA into the transgenic animal can be designed so as to produce transgenic animal models of human disease (Section 6.4).

References

1. Jeffreys, A.J. (1987) *Biochem. Soc. Trans.,* **15**, 309.
2. White T.J., Arnheim, N. and Erlich, H.A. (1989) *Trends Genetics,* **5**, 185.
3. Eisenstein, B.I. (1990). *New Engl. J. Med.,* **322**, 178.
4. Chelly, J., Concordet, J.-P., Kaplan, J.-C. and Kahn, A. (1989) *Proc. Natl Acad. Sci. USA,* **86**, 2617.
5. Newton, C.R., Graham, A., Heptinstall, L.E. *et al.* (1989) *Nucl. Acid Res.,* **17**, 2503.
6. Forrest, S. and Cotton, R.G.H. (1990) *Mol. Biol. Med.,* **7**, 451.
7. Burke, D.T., Carle, G.F. and Olson, M.V. (1987) *Science,* **236**, 806.
8. Olson, M.V. (1989) in *Genetic Engineering* Volume 11 (J.K. Setlow, ed.). Plenum Press, New York, p. 183.
9. Porteous, D.J. (1987) *Trends Genetics,* **3**, 177.

Further reading

Davies, K. (1988) *Genome Analysis: A Practical Approach.* IRL Press, Oxford.

Old, R.W. and Primrose, S.B. (1989) *Principles of Gene Manipulation*, 4th edn. Blackwell Scientific Publications, Oxford.

Sambrook, J., Fritsch, E.F. and Maniatis, T. (1989) *Molecular Cloning: A Laboratory Manual.* Cold Spring Harbor Laboratory Press, Cold Spring Harbor.

Williams, J.G. and Ceccarelli, A. (1992) *Genetic Engineering.* BIOS Scientific Publishers, Oxford.

4
MAPPING THE HUMAN GENOME

4.1 Genetic (meiotic) mapping

Genetic mapping depends on following the segregation of alleles at two or more loci during meiosis. Two loci A and B are genetically linked if the alleles present at those loci on a particular chromosome tend to be transmitted together through meiosis. To be linked, it is necessary but not sufficient for loci to be syntenic (that is, located on the same chromosome, e.g. chromosome 17). The combination of alleles at linked loci is called a haplotype: for example, haplotype A1B1 means a single chromosome carrying allele A1 at locus A and allele B1 at locus B. During meiosis, each pair of homologous chromosomes undergoes at least one recombination (cross-over) between non-sister chromatids. Thus to show genetic linkage, loci must be located in close physical proximity on a chromosome. For example, consider a chromosome on which loci A and B are close together but distant from a third locus C (*Figure 4.1*). A cross-over can occur at any position on the paired homologous chromosomes, but there is a much greater likelihood of cross-over in the B–C interval than there is in the A–B interval.

If there is a single cross-over in the B–C interval, the two non-recombinant products retain the original haplotypes A1B1C1 or A2B2C2. The two recombinant chromosomes show new A–C and B–C haplotypes (A2C1, B2C1, A1C2, B1C2). However, all four chromosomes maintain the original A–B haplotypes (A1B1 or A2B2). The great majority of meiotic products will have non-recombinant A–B haplotypes, consistent with genetic linkage between A and B.

4.1.1 Meiotic recombination and genetic map distance

During meiosis, recombination can be visualized cytogenetically as a chiasma. In human male meiosis, on average about 52 chiasmata are seen distributed over the 23 pairs of homologous chromosomes [1] (*Table 4.1*). Each pair of homologous chromosomes (bivalent) has at least one chiasma, and the number of chiasmata per bivalent is roughly proportional to the size of the chromosome, varying from about four on average on chromosome 1 to approximately one on average on chromosome 21. However, the location of cross-overs is not random: certain chromosomal areas are preferential sites of cross-overs (recombinational hotspots) whereas others show significantly reduced recombination frequency (Section 4.4.2). Although it is difficult

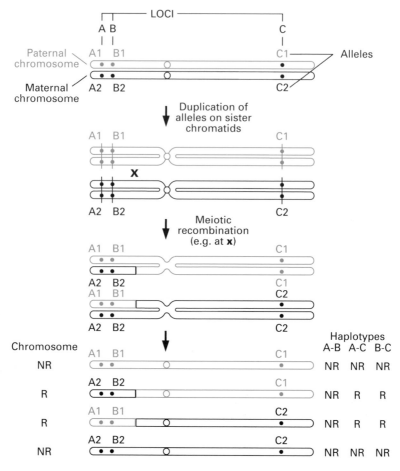

Figure 4.1: *Genetically linked loci are in close physical proximity on a chromosome.*

to count chiasmata in human female meiosis, genetic mapping of individual chromosomal regions suggests that crossing over occurs rather more frequently than in male meiosis (see below).

Each individual chiasma involves only two of the four chromatids in the bivalent. Double cross-overs can involve two, three or all four chromatids (*Figure 4.2*). As each of the two sister chromatids in a chromosome has only a 50% chance of being involved in any one chiasma, the maximum recombination fraction (i.e. recombination frequency) between any two loci is 0.5. The unit of genetic map distance, the Morgan, is defined as the length of chromosomal segment which, on average, undergoes one exchange per *individual* chromatid strand. Thus in the male, with an average of 52 chiasmata, the total map distance is about 26 Morgans. In female meiosis the number of chiasmata cannot be estimated easily, but linkage analysis suggests that the average map interval in female meiosis is approximately 85% longer than that in male meiosis (see Section 4.4.2). By extrapolation, therefore, the total female meiotic map distance is likely to be about 49 Morgans, and the sex-averaged total genetic map distance is likely to be about 37 Morgans, or 3700 centiMorgans (cM).

Table 4.1: *Mean chiasma frequency on individual bivalents in human male meiosis (adapted from reference 1)*

Bivalent chromosome number	Mean number of chiasmata	Bivalent chromosome number	Mean number of chiasmata
1	3.9	12	2.7
2	3.6	13	1.9
3	2.9	14	1.9
4	2.8	15	2.1
5	2.9	16	2.2
6	2.7	17	2.1
7	2.7	18	1.9
8	2.6	19	1.9
9	2.4	20	1.9
10	2.5	21	1.1
11	2.2	22	1.2

Over short chromosomal regions, the recombination fraction is directly proportional to the genetic map distance, so that a recombination fraction of 0.01 corresponds to a genetic map distance of 1 cM. However, over longer distances this linear relationship breaks down, mainly because of multiple cross-overs occurring between the two loci. Two loci which are non-syntenic (map to two different chromosomes) will show a recombination fraction of 0.5 and are not linked, either physically or genetically. Even syntenic loci may, if they are far enough apart on the chromosome, show a recombination fraction of 0.5 and not be genetically linked. Another factor contributing to the non-linear relationship between recombination fraction and genetic map distance is interference. Positive interference describes the effect that a cross-over has of reducing the probability of a second cross-over in its vicinity; this

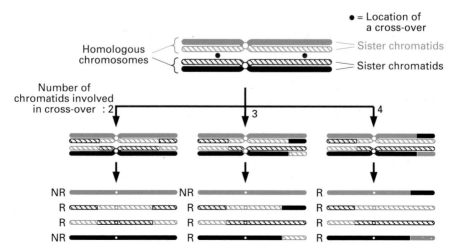

Figure 4.2: *Alternative outcomes of double cross-over. NR - non-recombinant chromosome. R - recombinant chromosome.*

appears to be limited to cross-overs occurring on the same chromosome arm. Various mapping functions have been proposed for relating the recombination fraction (θ) and the genetic map distance (w). A common mapping function for human work is the Kosambi function:

$$w \text{ (in cM)} = 1/4 \log [(1+2\theta)/(1-2\theta)].$$

Whichever mapping function is used, w increases exponentially as θ approaches 0.5 (*Figure 4.3*).

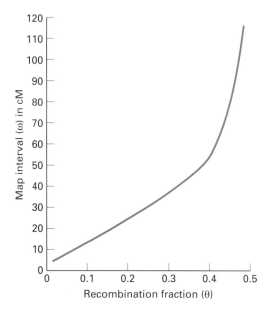

Figure 4.3: Relationship between genetic map distance (ω) and recombination fraction (θ) using Kosambi mapping function.

4.1.2 Markers for genetic linkage analysis

Conventional genetic linkage analysis uses pedigree data to check whether two loci are genetically linked and to estimate the recombination fraction between them. In human genetic mapping there are two common types of analysis:

(a) linkage between an unidentified disease gene (disease locus) and a previously mapped polymorphic marker locus, or a series of linked marker loci;
(b) linkage between two or more polymorphic loci which are known to be syntenic.

In either case, if a test locus is shown to be linked to a previously mapped marker, then the test locus must map close to the chromosomal location of the marker.

In order for a marker to be useful in genetic mapping, two requirements must be met:

(a) it must be able to detect polymorphism;
(b) its chromosomal, and preferably sub-chromosomal, location, must be known.

In the past, protein polymorphisms were the main markers used in linkage studies.

These included the highly polymorphic classical HLA antigens, blood groups and a variety of enzyme or structural protein polymorphisms. However, highly polymorphic protein markers are inevitably rare: all proteins are derived from a small percentage of the human genome which is under selective pressure to conserve coding DNA sequence and biological function. DNA polymorphisms are much more numerous, and now that we have the technology to study them, chromosome mapping has become much easier [2].

Most DNA markers define neutral polymorphisms in the large non-coding DNA component of the human genome. They include abundant conventional RSPs and also minisatellite and microsatellite VNTR-based polymorphisms. In order to screen for DNA markers which can recognize RSPs, isolated DNA clones are tested in turn to see if they can detect a frequent RFLP (Section 3.2.3). Each test involves hybridizing the DNA clone against a panel of Southern blots of human genomic DNA samples which have been digested with one of a variety of restriction nucleases (Section 3.2.3). Because the great majority of restriction sites are non-polymorphic, many restriction enzymes need to be tried. However, certain enzymes whose recognition sites include CpG, notably *Taq*I and *Msp*I, identify polymorphic sites relatively frequently because of the greater mutability of the CpG dinucleotide (Section 2.6).

Screening for microsatellite VNTR polymorphism involves hybridizing a chemically synthesized microsatellite probe to DNA from isolated DNA clones. Any microsatellite sequences identified are investigated further by sequencing the non-repetitive DNA flanking the microsatellite locus, then developing a PCR-based screen for the VNTR polymorphism (see Section 3.3.2).

The polymorphism information content (PIC) of a marker can range from 0 (always uninformative) to 1 (always informative). For a marker with n alleles,

$$\text{PIC} = 1 - \sum_{i=1}^{n} p_i^2 - \sum_{i=1}^{n-1} \sum_{j=i+1}^{n} 2p_i^2 p_j^2$$

where p_i is the frequency of the ith allele.

PIC values are highest for markers with several common alleles, such as VNTR markers. For example, the hypervariable VNTR locus in the 3' flanking region of the α-globin gene has about 30 alleles and a PIC value greater than 0.9. By contrast, the maximum possible PIC value for an RSP (RFLP) marker with two alleles is 0.5.

4.1.3 Isolation of chromosome-specific and chromosome band-specific DNA markers

The chromosomal location of individual DNA markers can be determined by various physical mapping methods (see Section 4.2). Formerly, DNA markers were usually random DNA clones which had been isolated from DNA libraries and subsequently mapped to a particular chromosome. Recently, more systematic DNA cloning approaches have permitted the isolation of markers from pre-determined chromosomal or subchromosomal locations.

Chromosome-specific DNA markers are isolated from chromosome-specific DNA libraries, in which the starting material is genomic DNA prepared from many copies of a single chromosome [3]. The different human chromosomes can be separated by flow-karyotyping. Chromosome preparations are stained with a DNA-binding dye, usually ethidium bromide, which can fluoresce in a laser beam. The amount of fluorescence exhibited by a given chromosome is proportional to the amount of dye bound, which in turn is largely proportional to the amount of DNA, and hence the size of the chromosome. Chromosomes can therefore be fractionated by size in a fluorescence-activated cell sorter. A stream of droplets containing stained chromosomes is passed through a laser beam at a rate of about 2000 chromosomes per second, and the fluoresecence of individual chromosomes is monitored by a photomultiplier. The resulting flow-karyotypes record the distribution of chromosomes in the sample (*Figure 4.4*). Droplets containing chromosomes of a particular size can be deflected on to particular co-ordinates of a collecting grid. DNA is then extracted from the fractionated individual chromosome sets and amplified by cell-based DNA cloning.

DNA markers from a particular subchromosomal region, such as a specific chromosomal band, can be obtained from chromosome microdissection DNA libraries. A micromanipulator with very fine needles is used to cut out a particular band from individual metaphase chromosomes under the microscope (*Figure 4.5*). When sufficient material has been collected (by dissecting the same region from several chromosome preparations), DNA is isolated and cloned using a PCR-based method [5]. Although the starting amount of DNA is tiny, respectable microdissection libraries

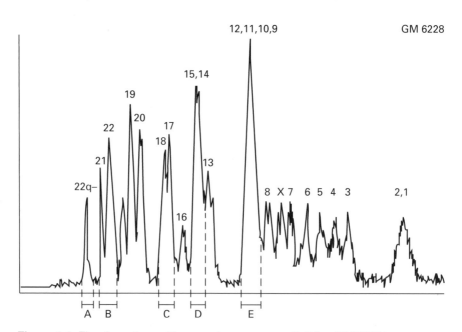

Figure 4.4: *Flow karyotype of human chromosomes. Cell line GM6228 has an unbalanced constitutional translocation t(11;22)(q23;q11). A–E: sorting windows used to collect specific sets of chromosomes or single chromosomes. Reproduced from reference 4 with permission from Academic Press.*

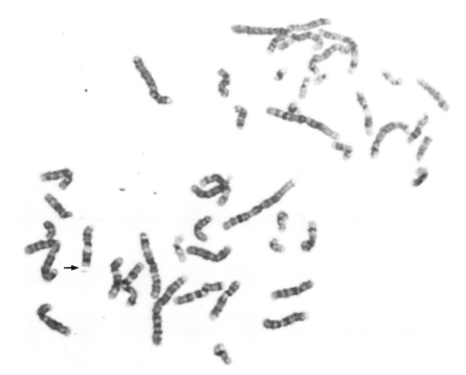

Figure 4.5: *Human chromosome microdissection. Arrow marks the position of a microdissected 11q24 band. Photograph courtesy of Debra Lillington, ICRF Department of Medical Oncology, St Bartholomew's Hospital, London.*

have been prepared from several chromosomal regions. The small microdissection DNA clones can be used as hybridization probes to identify large human DNA inserts in conventional genomic DNA libraries.

4.1.4 Pedigree-based genetic linkage analysis

In humans, unlike in experimental organisms, conventional linkage analysis is restricted to naturally occurring pedigrees. In such crosses the information needed to identify unambiguously where the cross-overs have occurred is generally lacking. To assess genetic linkage in humans, therefore, an indirect statistical approach is used which is based on maximum likelihood estimations. A likelihood ratio is established:

$$\frac{\text{likelihood the two loci are linked (recombination fraction} = \theta)}{\text{likelihood the two loci are unlinked (recombination fraction} = 0.5.}$$

For convenience, this ratio is normally expressed as a logarithm (base 10): the lod score (logarithm of the odds). Evidence for linkage is considered significant when the lod score (z) exceeds 3.0 (i.e. the likelihood ratio shown above exceeds 1000). A lod score of -2 or less is taken as evidence against linkage. If there are sufficient negative linkage data, a locus can be excluded from whole chromosomes, or even most of the genome (exclusion mapping). The search can then be focused on the remaining non-excluded locations. The apparently high likelihood ratio for proof of linkage is needed because the prior probability that any two loci are linked is about 1 in 50; the supplementary probability of 1 in 20 required to achieve a lod score of 3.0 therefore corresponds to only about 95% confidence that linkage exists.

In order to map an unidentified disease locus by genetic linkage, a panel of markers is tested in turn for evidence of co-segregation with the disease locus at meiosis. In the hypothetical example in *Figure 4.6* the disease is dominantly inherited, so that at the disease locus X each affected individual has one disease allele (X^D) and one normal allele (X^N). In all informative meioses, affected individuals transmit allele A2 with the disease (11 times) or allele A1 with the normal allele N (7 times). Allele A1 is never transmitted with the disease, nor allele A2 with the normal allele at the disease locus. The haplotypes are therefore A2–X^D and A1–X^N. No recombination has occurred between the X and A loci in 18 meioses, suggesting that the A locus is tightly linked to the disease locus X. By contrast, neither allele of the B marker preferentially segregates with the disease.

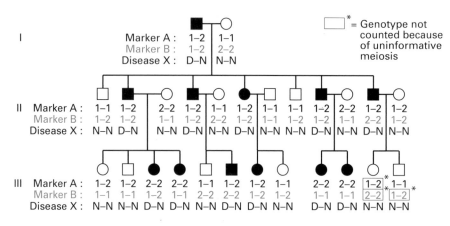

Transmission of haplotypes by affected individuals

Haplotype	No. of meioses	Haplotype	No. of meioses
A1–X^D	0	B1–X^D	5
A2–X^D	11	B2–X^D	6
A1–X^N	7	B1–X^N	4
A2–X^N	0	B2–X^N	2

Figure 4.6: *Testing for genetic linkage of markers to an unidentified dominant disease locus.*

A single pedigree rarely contains enough informative meioses to prove linkage, because families are usually small and markers are often insufficiently informative. Lod scores must usually be summed from data on several pedigrees to achieve statistical significance. Computer programs, such as LIPED, are available for calculating lod scores for different values of θ. Given significant evidence of linkage, the recombination fraction (θ) is taken as the value (θ̂) at which the lod score (z) is at a maximum (ẑ) (*Figure 4.7*). Enough human markers are now available to permit successful

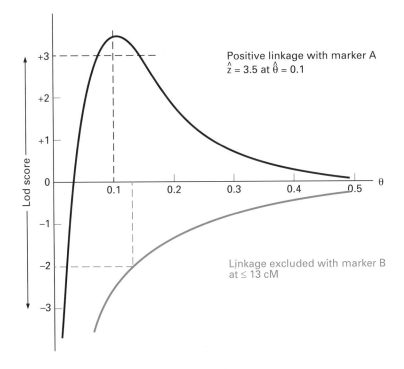

Figure 4.7: *Lod scores for possible linkage of a locus to two markers.*

mapping of any single gene disorder. Ideally one would use markers spaced about 20 cM apart throughout the whole genome. This would require about 200 markers; in practice, rather more are required because of uneven spacing between the available markers.

Multilocus linkage analysis is particularly valuable for assessing the most probable order of a series of linked loci. Additionally, it is employed to map the most probable location for an unidentified locus X (a disease gene, or unmapped marker locus), with respect to a framework of linked markers. The location score is defined as twice the natural logarithm of the likelihood ratio:

$$\frac{\text{likelihood of the data with X being inside the marker framework}}{\text{likelihood of the data with X being outside the marker framework}}$$

Sib-pair analysis is a form of linkage analysis used for mapping autosomal recessive disorders and for searching for susceptibility genes in diseases with non-Mendelian inheritance. Sibs share 50% of their genes on average through common genetic descent. However, sibs affected with the same genetic disease will be expected to have inherited the same disease alleles from their parents. Consequently, using panels of DNA markers which map to diverse locations throughout the genome, a large sample of affected sib pairs can be tested to see if they show a tendency to share one chromosomal segment more often than expected by chance. If they do, it is likely that a gene conferring susceptibility to the disease maps to this chromosome segment.

4.1.5 Determination of genetic map distance by single sperm typing

Single sperm typing is a new method of measuring recombination frequency, which should allow much higher resolution in genetic mapping than conventional pedigree linkage analysis (typically 0.01 cM compared to 1–2 cM). The method should be well suited to ordering tightly linked loci by assaying three loci at a time in sperm from triply heterozygous males. DNA sequences are analyzed in single meiotic products, that is, sperm cells [6]. Individual sperm cells are collected by micromanipulation or flow cytometry. DNA is extracted from single cells and typed to determine which alleles are present at the loci of interest. This is achieved by PCR amplification of the desired loci and hybridization of the PCR product with appropriate allele-specific oligonucleotides (see Section 3.2.4).

4.2 Low resolution physical mapping

4.2.1 Hybrid cell mapping

Somatic cell hybrids derive from experimentally induced fusion of cultured cells from different species. Hybrids used in human genetic mapping are made by fusion of a human cell and a rodent cell, usually from a mouse or hamster. Initial fusion products have a hybrid nucleus, a heterokaryon, which contains both human and rodent chromosomes. Such products are unstable and in subsequent rounds of cell division most human chromosomes fail to replicate and are lost. Stable fusion products are finally obtained which contain a complete set of the rodent chromosomes together with a few human chromosomes. The human chromosomes can be distinguished by their different morphology and differential staining with DNA-binding dyes, or they can be identified by screening for human DNA sequences or gene products which are known to map to specific chromosomes.

Human genes specifying recognizably human enzymes or other proteins can be mapped to a particular chromosome by testing individual hybrid clones for the presence of the human gene product. More generally, anonymous DNA clones can be assigned to a particular human chromosome by DNA hybridization or (if some of the clone's DNA sequence is known) by a PCR-based screen of a panel of hybrid cells.

Conventional somatic cell hybrids permit the construction of synteny maps, where panels of markers are mapped to particular chromosomes, but they do not provide information on the subchromosomal location of markers. This requires special hybrids which contain only part of a specific human chromosome. For example, translocation hybrids can be made from cells containing a chromosomal translocation, and deletion hybrids from cells with terminal or interstitial deletions; these permit assignment of DNA probes to subchromosomal segments (*Figure 4.8*).

A variant of the deletion hybrid approach involves using an existing somatic cell hybrid which contains a single human chromosome to generate a panel of deletion hybrids containing different small fragments of the chromosome. Controlled X-ray irradiation cleaves the chromosomes into comparatively small pieces essentially at random. The irradiated hybrid is fused with a rodent cell and cells containing human sequences are selected (e.g. by screening for the Alu repeat sequence). This results in a collection of irradiation-reduced hybrids, often called radiation hybrids. When DNA samples from a panel of such radiation hybrids are screened by hybridization or PCR

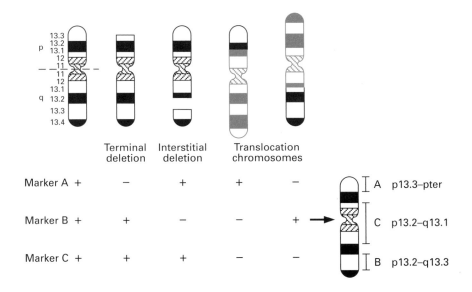

Figure 4.8: *Regional localization of chromosome 19 markers using a panel of deletion and translocation somatic cell hybrids.*

against a series of DNA clones, the patterns of cross-reactivity can be interpreted statistically to produce a linear map order for the DNA clones [7].

4.2.2 In-situ *hybridization*

Any purified DNA sequence can be assigned to its chromosomal location by labeling it and hybridizing it directly to the DNA of intact chromosomes. An air-dried microscope slide preparation of metaphase chromosomes is used, in which the chromosomal DNA has been denatured by exposure to formamide. Traditionally, *in-situ* hybridization has used ^{3}H-labeled DNA probes. Following autoradiography, positive signals are identified by counting silver grains in the slide emulsion and using a statistical test to discriminate genuine signal from background noise.

Recently, the sensitivity and resolution of in situ hybridization has been significantly increased by the development of fluorescence *in-situ* hybridization (FISH) [8]. In this technique, the DNA probe is labeled by addition of a reporter molecule. Following hybridization and washing to remove excess probe, the chromosome preparation is incubated in a solution containing a fluorescently labeled affinity molecule which binds to the reporter on the hybridized probe. To increase the intensity of the hybridization signal, large DNA probes are used, usually cosmid clones containing around 40 kb of insert. Because such large sequences will contain many interspersed repetitive DNA sequences, it is necessary to use competitive suppression hybridization. Before the main hybridization, the probe is mixed with an aqueous solution of unlabeled total genomic DNA. This saturates the repetitive elements in the probe, so that they no longer interfere with the specific *in-situ* hybridization of the unique sequences.

FISH has the advantage of providing rapid results which can be conveniently scored by eye using a fluorescence microscope. In metaphase spreads positive signals show as double spots, corresponding to probe hybridized to both sister chromatids (*Figure 4.9*). Using sophisticated equipment and reporter-binding molecules carrying different fluorophores, it is possible to map and order several DNA clones simultaneously. At present the maximum resolution of FISH on metaphase chromosomes is about 10 Mb. However, by using several differently labeled probes in interphase nuclei, much finer mapping is possible. Chromosomal locations cannot be identified in interphase preparations, but the relative order and spacing of different colored fluorescent probes can be established with a resolution as fine as 50 kb [9].

A special application of FISH has been the use of a whole chromosome-specific DNA probe (e.g. the combination of all human DNA inserts in a chromosome-specific DNA library) to hybridize to many loci spanning a whole chromosome. This causes whole chromosomes to fluoresce (chromosome painting) [10]. Painting of sub-chromosomal regions is also now possible, using DNA clones from a micro-dissected chromosomal region as a probe, and is likely to find applications in the study of chromosomal evolution and the definition of marker chromosomes in clinical and tumor cytogenetics.

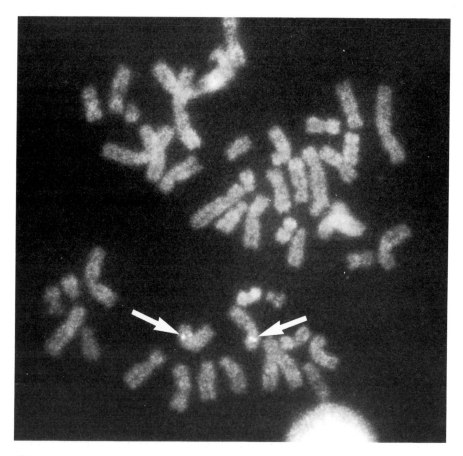

Figure 4.9: *Fluorescence* in-situ *hybridization. Mapping of a dystrophin cosmid clone against a human metaphase spread. Arrows denote location of positive hybridization signals.*

4.2.3 Hybridization to DNA from defined chromosomes or chromosomal regions

Individual types of human chromosomes can be purified by flow-karyotyping procedures (see Section 4.1.3). A probe can then be mapped simply by labeling it and hybridizing to dot-blots (see Section 3.1.6) which contain a panel of denatured DNA samples from the 24 different human chromosomes. Alternatively, if some of the DNA sequence is known, individual chromosomes can be tested for the presence of the sequence by a PCR assay (Section 3.3) using oligonucleotide primers derived from the sequence to be mapped.

More precise chromosomal localization can be obtained by analogous methods using DNA prepared from specific subchromosomal regions which have been excised by chromosome microdissection (Section 4.1.3).

4.2.4 Mapping by gene dosage

An enzyme encoded by an autosomal gene would be expected to show 50 or 150% of its normal activity if one or three gene copies were present, rather than the normal two. Consequently, cells which are monosomic or trisomic for a particular human chromosome (or subchromosomal region because of a translocation or deletion) can be assayed for differences in the quantity of a specific enzyme. Correlation between abnormal enzyme activity and abnormal chromosomal representation may therefore permit assignment of a map location for the gene encoding that enzyme.

4.2.5 Mapping of genes by defining associated chromosomal aberrations

Many human cancer genes have been mapped cytogenetically to a specific chromosome by demonstrating an increased frequency of particular chromosome aberrations in the relevant tumor cells. Tumor-specific chromosome breakpoints often lie near oncogenes (Section 5.4). Tumor-specific mutations which inactivate tumor suppressor genes often involve large-scale changes, such as loss of a whole chromosome or deletion of a large chromosome fragment containing the gene in question. Additionally, genes which are involved in other specific genetic disorders may occasionally be mapped by correlation with deletions, translocations and suchlike, which may occur in a small minority of patients.

4.2.6 Mapping of tumor suppressor genes by loss of constitutional heterozygosity

Large tumor-specific mutations which inactivate tumor suppressor genes can be identified at the molecular level by comparing DNA from a blood sample with DNA from a tumor sample from the same individual. DNA markers close to the tumor suppressor gene locus which are constitutionally heterozygous (e.g. in the blood DNA) are often hemizygous in the tumor DNA (*Figure 4.10*). This is a result of loss of one allele due to monosomy, terminal chromosome deletion or other genetic processes (see Section 5.4.3). This form of mapping can provide a rapid initial localization for a tumor suppressor gene (*Table 4.2*).

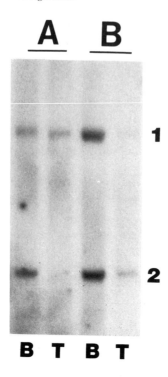

Figure 4.10: Loss of constitutional heterozygosity for the marker probe D22S15 in acoustic neuromas from two individuals. DNA samples are: B – blood DNA; T – tumor (acoustic neuroma) DNA. Sacl alleles are: 1 – 8.2 kb; 2 – 2.6 kb.

4.3 High resolution physical mapping

The physical mapping methods described above typically have a lower resolution limit of several megabases; they are complemented by molecular mapping methods which can map DNA in the range 1 bp to several megabases. DNA sequencing (Section 3.4) provides the ultimate physical map by determining the linear order of single nucleotides,

Table 4.2: Examples of allele losses in human tumors defined by DNA studies

Tumor	Site of allele loss
Retinoblastoma	13q
Osteosarcoma	13q
Wilms tumor	11p
Renal carcinoma	3p
Pheochromocytoma	1p, 22
Medullary thyroid carcinoma	1p
Rhabdomyosarcoma	11p
Meningioma	22
Acoustic neuroma	22
Breast	11p, 13q, 17p, 3p
Stomach	13q
Colon	5q, 17p, 18q, 22, others
Lung small cell	3p, 13q, 17p
Lung – others	3p especially

but mapping large DNA regions by sequencing is currently technically difficult (see below). Restriction mapping permits coarser molecular mapping over a range from about 0.1 kb to over 1 Mb (Sections 3.2.1 and 3.6.2).

4.3.1 Contig assembly, chromosome walking and chromosome jumping

Presently, the largest single human DNA fragments that can be mapped by molecular methods are DNA clones obtained in YACs (Section 3.6.1). These range in size up to 1 Mb. However, larger maps can be produced by identifying clones which have overlapping DNA inserts. The rationale of this approach is to assemble large contigs, chromosomal regions in which the entire DNA has been cloned as a series of overlapping fragments in identified clones. Genomic DNA libraries normally contain clones with overlapping inserts because they are made from many copies of each chromosome which have been cleaved almost at random (see *Figure 4.11*). When the individual DNA fragments are inserted into recipient cells during construction of the library, the overlapping fragments end up in separate cells.

Various strategies exist for identifying clones with overlapping inserts in a library. For example, specific short oligonucleotide probes can be hybridized to gridded arrays of isolated clones to test for sharing of specific sequences. Other methods of clone fingerprinting include restriction mapping (identifying a diagnostic pattern of restriction sites which is common to independent DNA clones), distribution of repetitive elements (e.g. *Alu* or *Kpn* repeats), and PCR amplification. In the latter case a short stretch of a DNA clone must first be sequenced in order to establish a sequence-tagged site (STS), from which primers can be designed which will amplify in all DNA clones which contain this site, but not in other clones.

Chromosome walking is another commonly used hybridization-based approach to expanding a map. An end fragment of a DNA clone is used as a hybridization probe to access other overlapping DNA clones (*Figure 4.11*). The range of chromosome walking techniques has recently been greatly improved by the use of YAC clones, which permit large walking steps. However, because of previous difficulties with chromosome walking, some attention has been devoted to the use of chromosome jumping. In this technique a DNA clone is used to isolate non-adjacent clones from the same chromosomal vicinity. To construct a jumping library, large fragments of genomic DNA are forced to circularize, and to include a foreign marker DNA sequence in forming the circle. The circularized DNA is digested with a restriction enzyme which does not cut within the marker sequence. This produces small fragments which are cloned, and clones containing the marker sequence are selected. The cloned fragments in a jumping library will contain two short human DNA sequences flanking the marker DNA on either side, representing sequences which were originally derived from distant chromosomal locations but which were artificially brought together by the cyclization reaction. Consequently it is possible to make chromosome 'jumps' of several hundred kilobases to move from one chromosomal location to another.

4.3.2 Identification of genes within cloned DNA

Once DNA from a desired chromosomal region has been cloned, coding sequences can be identified by a variety of methods.

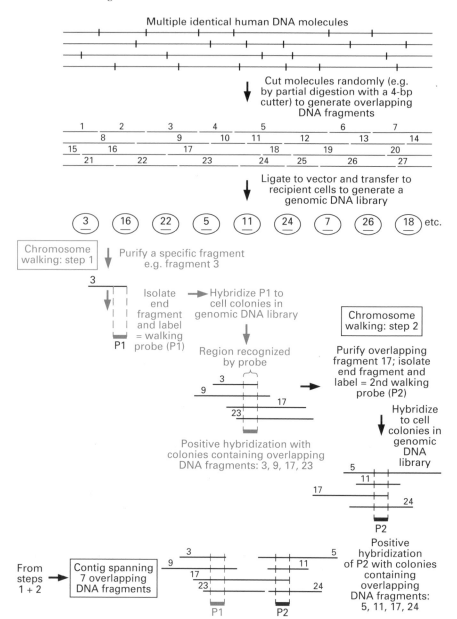

Figure 4.11: *Chromosome walking and contig assembly.*

Zoo-blots. This approach is based on the observation that coding sequences are strongly conserved during evolution, whereas non-coding DNA is not (see *Table 2.3*). A DNA clone which may contain a gene can be hybridized against a Southern blot of genomic DNA samples from a variety of animal species, a zoo-blot. At reduced hybridization stringency, probes containing human genes, but not probes containing only non-coding sequence, will generally show strong hybridization to animal genomic DNA. A small minority of genes are species-specific and will not be detected by this assay.

CpG islands/HTF islands. CpG islands are short stretches of DNA, often 1 kb or less, containing CpG dinucleotides which are unmethylated and which, unlike in the remainder of the genome (Section 2.6), are present at the expected frequency [11]. Such islands of normal CpG distribution are thought to mark transcriptionally active DNA sequences and are often found at the 5' end of housekeeping genes. CpG islands can be identified by diagnostic patterns of short- and long-range restriction maps using certain restriction nucleases whose recognition sites contain one or two CpGs. For example, *Hpa*II cleaves at the recognition site CCGG, but will not cleave CC^{m}GG. Regions containing methylated CpGs therefore have only infrequently occurring *Hpa*II sites. By comparison, *Hpa*II cleaves DNA in CpG islands much more frequently, so that the islands are also referred to as HTF islands (*Hpa*II tiny fragments). Rare-cutter enzymes such as *Bss*HII (GCGCGC), *Sac*II (CCGCGG) and *Not*I (GCGGCCGC) often cleave within CpG island sequences because the islands are relatively enriched in G + C and in CpG dinucleotides. Clustering of such sites is therefore commonly associated with CpG islands.

Hybridization patterns with mRNA/cDNA. A direct method for detecting a housekeeping gene is to hybridize the candidate DNA against a Northern blot of mRNA from blood lymphocytes (or a convenient tissue culture cell line), or against a cDNA library prepared from such. Genes with unidentified tissue specificity, however, may not always be identified even after screening against mRNA or cDNA from several different tissues.

DNA sequencing. DNA sequencing of a candidate region may identify comparatively long open reading frames, regions of DNA which could conceivably specify polypeptide sequence. Unlike non-coding DNA, polypeptide-encoding DNA sequences are under selection pressure to avoid mutations which introduce termination codons. Analysis of DNA sequencing data may also reveal certain DNA sequence motifs which are consistent with an expressed DNA sequence. Finally, comparison with all other known DNA sequences stored on computer databases may suggest something about the nature of the putative gene product, such as whether it is likely to be a specific type of gene (e.g. an actin gene), or whether it encodes a membrane protein or suchlike. Because large-scale DNA sequencing is rather laborious, however, it is usual to reserve this approach for the final analysis of an expressed DNA sequence.

4.4 The human genetic map

4.4.1 Current progress

Progress in human gene mapping is reviewed at successive international Human Gene Mapping Workshops. By mid-1990, over 6000 loci, of which only a minority are gene loci, had been mapped to specific chromosomal locations – see *Table 4.3*. Human Gene Mapping Workshop symbols are allocated to genes and pseudogenes, usually using two to six characters. For anonymous DNA sequences, the convention is to use

D (= DNA) followed by 1 to 22, X or Y to denote the chromosomal location, then S for a unique segment, Z for a chromosome-specific repetitive DNA family or F for a multi-locus DNA family, and finally a serial number (*Table 4.4*). Somatic cell hybridization and *in-situ* hybridization have been the most widely used techniques for assigning map positions to autosomal loci (*Table 4.5*). In the future, FISH (Section 4.2.2) is likely to be very widely used because of its high mapping resolution and relative simplicity.

The map of markers for human genetic linkage is rapidly improving. The first DNA polymorphism map to cover the genome was published in 1987 and had an average spacing between markers of 10–15 cM [13]. Recent linkage analyses have been facilitated by the international CEPH collaboration, by which DNA samples from international reference pedigrees chosen for excellent structure are available for analysis through the auspices of the Centre d'Etude du Polymorphisme Humain in Paris. By mid-1990, over 2000 different polymorphic loci had been recorded, permitting in theory an average marker spacing of 2 cM. However, recently published maps on individual chromosomes have map resolutions in the 5–10 cM range because in each case the linkage data have been obtained with a limited range of markers.

Table 4.3: *Mapped loci in the human genome (adapted from reference 12)*

Chromosome	Total number of loci	Number of gene loci (number of sequenced loci)		Number of polymorphic loci
1	311	192	(82)	146
2	196	116	(50)	90
3	786	75	(29)	130
4	242	73	(34)	138
5	192	74	(28)	112
6	207	110	(55)	86
7	555	121	(50)	189
8	172	58	(25)	55
9	110	65	(24)	47
10	156	62	(28)	88
11	624	140	(55)	189
12	155	103	(45)	56
13	122	29	(12)	53
14	98	56	(33)	51
15	126	52	(20)	49
16	335	59	(25)	122
17	451	99	(47)	150
18	55	23	(10)	32
19	194	82	(38)	59
20	64	37	(17)	22
21	202	34	(7)	60
22	238	57	(22)	99
X	730	179	(31)	235
Y	231	13	(5)	17

Table 4.4: Human gene mapping nomenclature

Symbol	Interpretation
CRYB1	Gene for crystallin, beta polypeptide 1
GAPD	Gene for glyceraldehyde-3-phosphate dehydrogenase (GAPD)
GAPDL7	GAPD-like gene 7, functional status unknown
GAPDP1	GAPD pseudogene 1
AK1	Gene for adenylate kinase, locus 1
AK2	Gene for adenylate kinase, locus 2
PGK1*2	Second allele at *PGK1* locus
DYS29	Unique DNA segment number 29 on the Y chromosome
D11Z3	Chromosome 11-specific repetitive DNA family number 3
DXYS6X	DNA segment found on the X chromosome, with a known homolog on the Y chromosome, and representing the 6th XY homolog pair to be classified
DXYS44Y	DNA segment found on the Y chromosome, with a known homolog on the X chromosome, 44th XY homolog pair
D12F3S1	DNA segment on chromosome 12, first member of multilocus family 3 (other members on X, 18, 21)
DXF3S2	DNA segment on chromosome X, second member of multilocus family 3

4.4.2 Relationship between physical and genetic maps

The sex-averaged total genetic map distance for the 3000 Mb human genome is about 3700 cM. Therefore, on average, a genetic map distance of 1 cM corresponds approximately to a physical map distance of 0.8 Mb. However, the ratios of genetic and physical map distances on individual chromosomal segments often deviate considerably from this average figure due to non-random location of chiasmata. Chromosomal segments containing recombinational hotspots will show a high cross-over frequency. One such hotspot is the approximately 2.5-Mb long pseudo-autosomal region of the Y chromosome, which undergoes an obligatory cross-over with the X chromosome during meiosis (see Section 2.2). The remainder of the Y chromosome does not

Table 4.5: Methods used to map autosomal loci

Method	Number of loci mapped by 1 March, 1990
Somatic cell hybridization	1080
In-situ hybridization	623
Family linkage study	444
Dosage effect	156
Restriction enzyme fine mapping	150
Chromosome aberrations	117
Homology of synteny	93
Radiation-induced gene segregation	18
Others	138
Total (many mapped by two or more methods)	2819

undergo recombination. In general, there is a high recombination frequency at telomeres, while recombination is suppressed near to centromeres and, to a lesser extent, in sub-telomeric regions.

Longer genetic map distances in females suggest a higher cross-over frequency in female than in male meiosis. Maps reported for 14 of the 22 autosomes at Human Gene Mapping Workshop HGM10.5 show the female genetic map distance was on average 1.86 times greater than the equivalent male map distance. The female–male ratio ranged from 1.21 (chromosome 14) to 2.75 (chromosome 6) (*Table 4.6*). However, certain chromosome segments show a larger genetic map distance in males than in females. These include segments of 11p and 11q. Neighboring chromosomal regions can occasionally show enormous differences in the ratio of female to male genetic map distances: for example, the distance between markers RH and L56 on 1p is 14 cM in male meiosis and only 0.2 cM in female meiosis, while in a neighboring chromosomal region the corresponding map interval between the C52 and L1039 markers is 0.1 cM in males and 28 cM in females (*Figure 4.12*).

4.4.3 Chromosomal organization of functionally related genes

Although the current gene map is far from complete, several inferences can be drawn concerning the chromosomal organization of functionally related non-allelic genes. Non-allelic genes which encode the same type of product are often clustered, whereas genes which encode protein isoforms restricted to specific tissues or subcellular compartments are often non-syntenic, as are genes encoding many other different types of functionally related products (*Table 4.7*).

Table 4.6: *Differences in human male and female linkage maps*

Chromosome number	Terminal markers of linkage map		Map distance (cM)	
	p	q	Male	Female
1	D1Z2	D1S68	239	392
3	D3S22	D3S26	171	224
4	D4S115	D4S119	127	287
5	D5S10	D5S43	217	377
6	F13A	D6S21	101	278
10	D10S31	D10S6	148	236
12	F8VWF	PAH	93	164
14	D14S26	D14S20	87	109
15	D15S24	D15S3	85	166
16	D16S85	D16S7	105	169
17	D17S34	D17S24	105	258
19	D19S21	PRKCG	81	156
20	D20S18	D20S19	53	137
21	D21S13E	COL6A1	73	103

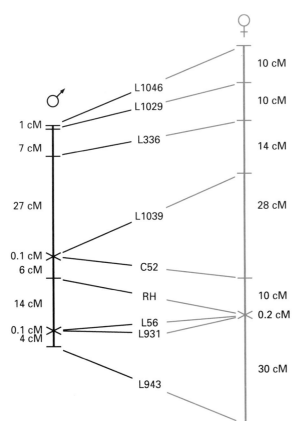

Figure 4.12: *Sex-specific differences in the genetic map for human chromosomal region 1pter. Data redrawn from reference 13.*

4.4.4 Morbid anatomy of the human genome

Many inherited single gene disorders which map to the X chromosome have long been recognized as such because of distinctive pedigree patterns. However, mapping a disorder to a specific subchromosomal region on the X chromosome, or to a specific autosome, has mostly been a phenomenon of the last decade, largely as a result of the rapid increase in the number of DNA markers available for linkage analysis. At the time of writing a considerable number of important inherited single gene disorders had been mapped (*Table 4.8*). In addition, many genes which predispose to somatic cancer mutations have been mapped (Section 5.4), as have genetic factors which contribute to common diseases such as atherosclerosis, diabetes and suchlike (Section 6.1).

Not surprisingly, because of the ease of identifying X-linked inheritance, more pathological disorders have been mapped to the X chromosome than to any other. Other chromosomes which currently have a comparatively high density of mapped disorders in relation to their size are chromosomes 11 (see *Figure 4.13*), 17, 19 and 22. In contrast, chromosomes 2, 5 and 18 have relatively few mapped disorders, as inevitably does the Y chromosome because of its paucity of functional genes. As expected from the foregoing section, disorders with very similar phenotypes can map to several different chromosomes. For example, in different families adrenal hyperplasia maps to chromosomes 1p, 6p, 8q, 10 and 15 because of genetic deficiencies in a variety of related steroidogenic enzymes (*Table 4.7*).

Table 4.7: *Distribution of genes encoding functionally related products*

Genes which encode	Organization	Examples
The same product	Often clustered	Genes encoding rDNA, histones, HLA, immunoglobulins
Tissue-specific protein isoforms or isozymes	Sometimes clustered; sometimes non-syntenic	Clustering of pancreatic and salivary amylase genes (1p21); non-synteny of α-actin genes expressed in skeletal (1p) and cardiac (15q) muscle
Isozymes specific for different subcellular compartments	Often non-syntenic	Cytoplasmic and mitochondrial isozymes of aldehyde dehydrogenase (*ALDH1* – 9q, *ALDH2* – 12q), aconitase (*ACO1* – 9p; *ACO2* – 22q), thymidine kinase (*TK1* – 17q; *TK2* – 16), etc.
Enzymes in the same metabolic pathway	Often non-syntenic	Genes for steroidogenic enzymes e.g. steroid 11-hydroxylase – 8q steroid 17-hydroxylase – 10 steroid 21-hydroxylase – 6p
Sub-units of the same protein or enzyme	Often non-syntenic	Hemoglobin: α – 16p; β – 11p; collagen: α(1)I – 7q; α(2)I – 17q ferritin: H – 11q; L – 22q class I HLA: H – 6p; L – 15q immunoglobulins: H – 14q; L – 2p or 22q
Ligand plus associated receptor	Often non-syntenic	Genes encoding interferons (*IFNA, IFNB* – both 9p, *IFNG* – 12q) and their receptors (*IFNAR, IFNBR,* – both 21q, *IFNGR* – 18); insulin gene *INS* – 11p, but insulin receptor *INSR* – 19p.

4.4.5 Comparative mapping

Comparison of the organization of the human genetic map with the genetic maps of other mammalian genomes is useful because:

(a) it permits insights into genome evolution;

(b) it facilitates assignment of map locations for unidentified human genes;

(c) it suggests specific animal mutants as candidates for animal models of human genetic disorders (Section 6.4).

Presently, the mouse genome is the only other mammalian genome to have been mapped in detail. Although both the human and mouse genomes contain approximately 3×10^9 bp of DNA, their chromosomal organizations are rather different (see Section 2.2) and the total haploid genetic length of the mouse genome is estimated to be 1600 cM, reflecting a lower recombination frequency than in man. Mapping of the mouse genome has been greatly facilitated by interspecific mouse back-crosses and more than 600 loci have been mapped with an average map interval of less than 3 cM [14].

Table 4.8: Map locations of some inherited single gene disorders

Disorder	Chromosomal location	Gene locus
α-thalassemia	16p13.3	α-globin
β-thalassemia	11p15.5	β-globin
Sickle cell disease	11p15.5	β-globin
Familial hypercholesterolemia	19p13	Low density lipoprotein receptor
Duchenne/Becker muscular dystrophy	Xp21	Dystrophin
Cystic fibrosis	7q31-q32	*CFTR*
Huntington's disease	4pl6.3	?
Fragile X-linked mental retardation	Xq27.3	*FMR-1*
Neurofibromatosis type I	17q11.2	*NF1*
Myotonic dystrophy	19q13	?
Spinal muscular atrophy	5q11.2-q13.31	?
Adult polycystic kidney disease	16p13.3	?
21-hydroxylase deficiency	6p21.3	21-hydroxylase
Hereditary hemochromatosis	6p21.3	?
Phenylketonuria	12q22-q24	Phenylalanine hydroxylase
α$_1$-antitrypsin deficiency	14q32.1	α$_1$-antitrypsin
Retinoblastoma	13q14.2	*RB1*
Polyposis coli	5q21	*APC*
Marfan syndrome	15q21.1	Fibrillin
Hemophilia A	Xq28	Factor VIII

Of the thousands of loci that have been mapped in mice and humans, more than 400 have been mapped in both species and are known to be homologs. Homologs of human X-linked genes are always located on the murine X chromosome. However, autosomal loci which are syntenic but loosely linked in one of the species are often non-syntenic in the other species [15]. For example, the *src* and *abl* oncogenes, and genes encoding β$_2$-microglobulin (*B2m*), follicle-stimulating hormone beta polypeptide (*Fshb*), and glucagon (*gcg*), are all encoded on mouse chromosome 2 but have homologs on human 20q, 9q, 15q, 11p and 2q, respectively. Nevertheless, tightly linked loci in one species are often syntenic in the other species (*Figure 4.14*).

The existence of short chromosomal regions which demonstrate homology of synteny has permitted inference of map positions for a number of human genes by comparison with the mouse map (see *Table 4.5*). An important application for this method of mapping is likely to be in mapping complex multifactorial diseases for which an animal model exists. In humans such diseases are difficult to study because of clinical and genetic heterogeneity, and also partly because of the limited availability and informativeness of suitable individuals. However, mouse genetic mapping is relatively simple, given the flexibility in designing breeding crosses and the availability of a large number of highly informative murine microsatellite VNTR polymorphic

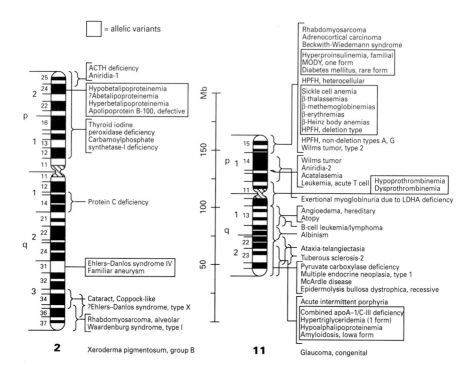

Figure 4.13: *Diseases which are known to map to chromosomes 2 and 11. Allelic variants are boxed. Redrawn from McKusick, V.A. (1990)* Mendelian Inheritance in Man, *9th Edn, with permission from the Johns Hopkins University Press.*

markers. Recently, mapping approaches have resulted in the identification of three genes, *Idd-3*, *Idd-4* and *Idd-5*, which confer susceptibility to insulin-dependent diabetes in the non-obese mouse model of diabetes (see Section 6.1.4)

4.5 The Human Genome Project

Serious consideration of a project to obtain the complete DNA sequence of the human genome began in the mid-1980s, initially by the US Department of Energy and shortly afterwards by the US National Institutes of Health. The resultant Human Genome Project, involving both centers, envisages completion of the project by the year 2005 at a cost of about $3 billion, approximately one-tenth of the cost of the Space Station. Additionally, various research centers in Europe and Japan are contributing efforts in the same direction and the Human Genome Organization (HUGO) has been established to co-ordinate international contributions to the program.

Because of limitations in current DNA sequencing technology, much of the initial effort is being directed to constructing improved genetic and physical maps of the human genome and in developing superior DNA sequencing technologies. The first

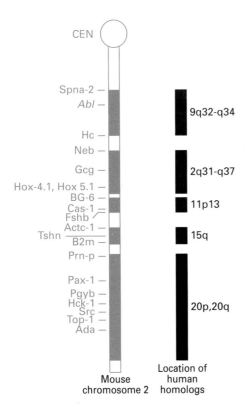

Figure 4.14: *Homology of synteny in human and mouse genetic maps. Human homologs of loci on mouse chromosome 2.*

5-year plan envisages a genetic map with markers spaced on average every 2 cM. The projected physical map would have sequence-tagged sites (Section 4.3.1) located every 0.1 Mb, and 2-Mb contigs of overlapping cloned DNA fragments covering large parts of the genome [16]. The major problem is the sheer size of the human genome; the total of 3000 Mb is more than 10 000 times the size of the biggest genome that has been sequenced to date (cytomegalovirus). Because of the problem of scale, several centers are concentrating initially on the expressed component of the genome, the 2–3% of the genome which is coding DNA. Because the detection of unidentified gene sequences in cloned genomic DNA is still not straightforward, as a first step partial DNA sequencing is being conducted on DNA clones from different human cDNA libraries, for example brain cDNA libraries [17]. The DNA sequence obtained is used to generate expressed sequence tags, which can in turn permit isolation of unidentified human genes.

Parallel with the human genome mapping endeavor, smaller projects have been established to sequence the genomes of model organisms such as *E. coli*, yeast, *Drosophila*, mouse and the plant *Arabidopsis*. It is hoped that sequencing of the smaller genomes may serve as pilot programs for large-scale sequencing of the human genome. Additionally, as genes are often conserved between species, the comparatively easy isolation of a gene from a relatively simple genome may permit a short cut to isolating a human homolog. In the long term, the additional information on well-studied organisms is likely to be of great importance in understanding fundamental developmental processes.

As the Human Genome Project gathers momentum, an increasing concern is the collection and interpretation of the vast amount of data that are being generated. Presently, the Genome Database at the Johns Hopkin University, Baltimore, USA is acting as the main repository of mapping data while there are major established protein and DNA sequence databases in the USA, Europe and Japan.

References

1. Hulten, M. (1974) *Hereditas, 76*, 55.
2. White, R. and Lalouel, J.-M. (1988) *Sci. American, 258*, 20.
3. Davies, K., Young, B.D., Elles, R.G., Hill, M.E. and Williamson, R. (1981) *Nature, 293*, 374.
4. Cotter, F., Nasipuri, S., Lam, G. and Young, B.D. (1989) *Genomics, 5*, 470.
5. Ludecke, H.J., Senger, G., Claussen, U. and Horsthemke, B. (1989) *Nature, 338*, 348.
6. Cui, X., Li, H., Goradia, T.M. *et al.* (1989) *Proc. Natl Acad. Sci. USA, 86*, 9389.
7. Cox, D.R., Burmeister, M., Proce, E.R., Kim, S. and Myers, R.M. (1990) *Science, 250*, 245.
8. Lichter, P., Tang, C.-J.C., Call, K., Hermanson, G., Evans, G.A., Housman, D. and Ward, D.C. (1990) *Science, 247*, 64.
9. Trask, B.J., Massa, H., Kenwrick, S. and Gitschier, J. (1991) *Am. J. Hum. Genet., 48*, 1.
10. Lichter, P., Cremer, T., Borden, J., Manuelidis, L. and Ward, D.C. (1988) *Hum. Genet., 80*, 224.
11. Bird, A.P. (1987) *Trends Genetics, 3*, 342.
12. Stephens, J.C., Cavanaugh, M.L., Gradie, M.I., Mador, M.L. and Kidd, K. (1990) *Science, 250*, 237.
13. Donis-Keller, H., Green, P., Helms, C. *et al.* (1987) *Cell, 51*, 319.
14. Copeland, N.G. and Jenkins, N.A. (1991) *Trends Genetics, 7*, 113.
15. Nadeau, J.H. (1989) *Trends Genetics, 5*, 82.
16. Watson, J.D. (1990) *Science, 248*, 44.
17. Adams, M.D., Kelley, J.M., Gocayne, J.D. *et al.* (1991) *Science, 252*, 1651.

Further reading

Conneally, P.M. and Rivas, M.L. (1980) Linkage analysis in man. *Adv. Hum. Genet., 10*, 209.

Human Gene Mapping 10 (1989) Tenth International Workshop on Human Gene Mapping. *Cytogenet. Cell Genet., 51*.

Human Gene Mapping 10.5 (1990) Update to the Tenth International Workshop on Human Gene Mapping. *Cytogenet. Cell Genet., 55*.

McKusick, V.A. (1990) *Mendelian Inheritance in Man*, 9th Edn. Johns Hopkins University Press, Baltimore.

Various authors (1986) The Molecular Biology of *Homo sapiens. Cold Spring Harbor Symposiums in Quantitative Biology*, Volume 51. Cold Spring Harbor Laboratory Press, New York.

5
HUMAN DISEASE GENES: ISOLATION AND MOLECULAR PATHOLOGY

5.1 Isolation of human disease genes

5.1.1 Approaches used to isolate and identify human disease genes

The human disease genes that have been isolated to date are largely genes which are responsible for inherited single gene disorders or somatic cancers. Three major strategies have been employed:

(a) gene isolation via an expression product, an approach sometimes described as forward genetics;

(b) a candidate gene approach in which selected human genes that have previously been isolated are investigated in turn by looking for patient-specific mutations which would be expected to disrupt normal gene expression;

(c) a DNA mapping and cloning strategy based simply on knowledge of the disease gene's chromosomal location, sometimes referred to as reverse genetics or positional cloning.

The 'forward genetics' approach. This was the first approach to be developed. In the case of diseases where the biochemical basis is known, characterization of the gene product can permit methods of isolating the gene. Normally, the first step is to isolate a cDNA clone corresponding to the known protein product. Thereafter the specific cDNA clone can be hybridized against a human genomic DNA library to isolate normal human gene clones. These in turn can be used to isolate disease alleles either by screening patient-specific DNA libraries, or more conveniently, by PCR amplification from patient DNA samples.

Following partial purification of the gene product, the initial isolation of a specific DNA clone can be achieved by a variety of methods. For example, phenylketonuria is due to a defect of the enzyme phenylalanine hydroxylase. Following purification of this enzyme, specific antibodies were raised. The antibodies were employed in turn to identify specific mRNA corresponding to the protein product in *in-vitro* protein synthesis assays of rat liver polysomal mRNA. The purified mRNA was then converted to cDNA and a specific cDNA clone isolated. More recently, direct antibody screening has become possible using specialized cDNA expression libraries.

An alternative first step in the forward genetics approach is to obtain some amino acid sequence from the protein product. This permits the design of a panel of synthetic oligonucleotides corresponding to all possible codon permutations which can be translated to give the established amino acid sequence. The panel of oligonucleotides can then be hybridized to a human cDNA (or even genomic DNA library) in order to isolate a cognate DNA clone. This was the approach used to isolate the first factor VIII DNA clones; microsequencing of porcine factor VIII was followed by synthesis of a panel of corresponding oligonucleotides and successful screening of a porcine genomic DNA library. The porcine factor VIII DNA clone was then used to screen a human genomic DNA library.

The candidate gene approach. As more and more human genes are isolated, and techniques for identifying single base mismatches become more sophisticated (Section 3.5), the candidate gene approach is becoming more widely used. Previously isolated human genes can be considered as disease locus candidates if they are known to have a role in the physiology of the diseased tissue or if they are known to map to the same chromosomal area and encode a plausible protein. For example, in the case of hereditary retinal degeneration, numerous genes encoding proteins involved in phototransduction have already been cloned. Investigation of one of these, a rhodopsin gene, has led to the identification of mutations in some patients with autosomal dominant retinitis pigmentosa [1]. In the case of Marfan Syndrome, initial linkage analyses mapped the gene to 15q and were followed by localization of the connective tissue protein, fibrillin, to 15q21.1 by *in-situ* hybridization. Investigation of the structure of the fibrillin gene in normal individuals and Marfan patients led to the identification of a mis-sense mutation, suggesting that fibrillin is the Marfan Syndrome disease locus [2].

The positional cloning ("reverse genetics") approach. If nothing is known about the nature of the disease gene product and no likely candidate disease genes can be identified, it is still possible to isolate an unidentified disease gene by the positional cloning approach. Recently, this approach has been outstandingly successful; several important human disease genes whose biochemical bases were not known, were isolated shortly after their chromosomal locations were established (*Table 5.1*).

In each case the starting point in such approaches has been to establish a subchromosomal localization for the disease gene. Often this is achieved by linkage analysis, and in the special case of tumor suppressor genes, by loss of constitutional heterozygosity in tumors. Subsequent linkage analyses can then be employed to identify flanking markers that show very little or no recombination with the disease locus. Following this stage a variety of molecular techniques (including chromosome jumping and walking, assembly of YAC and cosmid contigs, etc. – see Section 4.3.1) can be used to proceed from the marker loci towards the disease gene locus. In some cases the approach to identifying the disease gene is helped enormously by the identification of disease-specific chromosomal deletions or translocations. Otherwise the route to the gene involves saturation genetic and physical mapping. The ultimate identification of a disease locus minimally requires the demonstration of patient-specific mutations which are inconsistent with normal gene expression.

Table 5.1: *Isolation of human disease genes by positional cloning*

Disease gene	First reports of chromosomal localization	First report of gene cloning and isolation
Duchenne/Becker muscular dystrophy (dystrophin)	1979–1982: Xp21–autosome translocations 1982: linkage analysis	1985–1986
Chronic granulomatous disease	1984–1985: Xp21 deletion patients	1986
Retinoblastoma	1983: deletion and linkage analyses → 13q14	1986–1987
Wilms tumor	1978–1979: identification of deletions → 11p13	1990
Cystic fibrosis	1985: linkage analyses → 7q	1989
NF1	1987: linkage analyses → 17q 1989: patients with17q11.2 translocations	1990
Choroideremia	1985: linkage analyses → Xq13–Xq21	1990
Colorectal cancer		
DCC gene	1988: deletion analysis → gene maps to 18q21–qter	1990
APC gene (adenomatous polyposis coli)	1986: cytogenetic identification of deletion 1987: linkage analyses → 5q21	1991
Fragile X-associated mental retardation	Cytogenetic analyses → Xq27	1991

5.1.2 Isolation of disease genes via a chromosomal deletion

The identification of a small chromosomal deletion associated with a disease can afford a rapid route to isolating the disease gene. For example, genomic DNA clones representing part of the human dystrophin gene were first isolated following investigation of a boy who suffered from Duchenne muscular dystrophy (DMD), chronic granulomatous disease and retinitis pigmentosa [3]. This patient had a contiguous gene syndrome due to a single large deletion which removed DNA sequence from neighboring genes at a single chromosomal location, Xp21. As the deletion on this boy's X chromosome was presumed to include the gene responsible for DMD, a method of subtraction cloning was devised to prepare a small subchromosomal gene library of DNA sequences corresponding to the deleted region (*Figure 5.1*).

Genomic DNA from a normal individual was digested with a restriction nuclease, *Mbo*I which cleaves DNA immediately 5' of its recognition sequence GATC. This produced fragments which were double-stranded except for a single-stranded sequence GATC at the extreme 5' ends, that is, unpaired 5' GATC overhangs. The *Mbo*I fragments were denatured and combined with a 200-fold excess of denatured DNA from the deletion patient, which had been sonicated to produce ragged ends. Subsequently, a vector cut with *Bam*HI was used to clone annealed fragments. As this vector could only accept fragments with overhanging 5' GATC ends on *both* DNA strands, Xp21

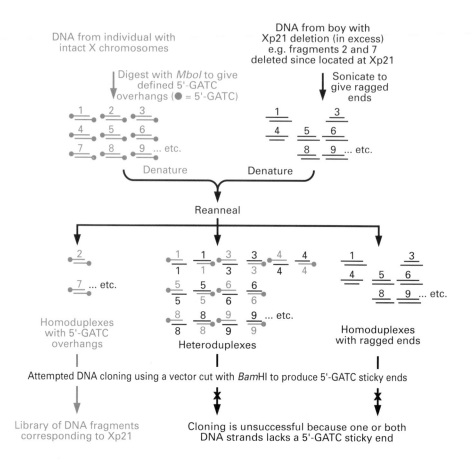

Figure 5.1: *Subtraction cloning.*

sequences were preferentially cloned (the other fragments with 5' GATC ends from the normal individual formed heteroduplexes with ragged end fragments from the patient DNA). Individual DNA clones in the resulting subtraction library were then used as probes in Southern blot hybridization against normal and DMD patient DNA samples. One DNA clone was tightly linked to the disease in family studies and detected DNA deletions in a proportion of DMD patients. This was the first of a series of dystrophin gene clones to be isolated.

The above strategy was helped by the fact that the boy with the deletion was hemizygous; the presence of a single X chromosome meant that there was no homologous chromosome carrying sequences corresponding to the deleted region. In the case of patients with small deletions on a single autosome, it is first necessary to isolate the deletion chromosome in a somatic cell hybrid so that the subtraction cloning strategy is not defeated by the presence of sequences on the homologous chromosome which correspond to the deleted region.

5.1.3 Isolation of disease genes via a chromosomal translocation

Characterization of disease-associated translocation breakpoints has afforded direct molecular access to a number of disease genes, including the recently identified gene for neurofibromatosis type I (NF1). Two NF1 patients were observed to have a chromosomal translocation involving chromosome 17q and another chromosome. In each case the translocation had involved breakage of chromosome 17 at a position within the *NF1* gene itself, thereby disrupting the expression of the *NF1* gene (*Figure 5.2*). In one approach to identify the *NF1* gene, a chromosome 17-specific DNA library was used, in which the DNA clones were selected to contain *Not*I sites, a *Not*I linking library. Rare-cutter restriction sites often mark transcriptionally active DNA sequences (Section 3.6.2) and the *Not*I linking library was therefore expected to have a high density of chromosome 17 gene clones. Individual DNA clones from the library were used as hybridization probes against total genomic DNA from the translocation patients and normal individuals which had been digested with rare-cutter restriction nucleases and size-fractionated by PFGE. One such clone, 17L1A, identified anomalous-sized restriction fragments in the DNA from the t(1;17) translocation patient, suggesting that it mapped in the immediate vicinity of the translocation [4] (*Figure 5.2*). Exhaustive cloning and mapping of this region led to the identification of a number of genes, one of which could be identified as the *NF1* gene by the demonstration of patient-specific mutations.

5.1.4 Isolation of disease genes by saturation genetic and physical mapping

In some single gene disorders extensive analysis of patients has failed to identify evidence of large mutations which could suggest molecular strategies for obtaining gene clones. In such cases it has been possible to isolate the disease gene after extensive genetic mapping and exhaustive molecular characterization of the chromosomal

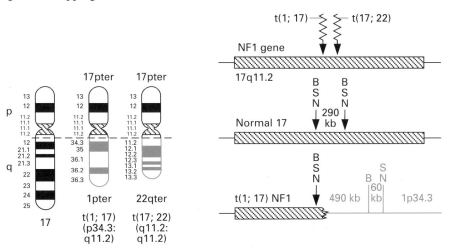

Figure 5.2: *Disease-associated translocation breakpoints and associated long-range restriction maps in the NF1 gene. Rare-cutting restriction enzyme sites are: B, BssHII; N, NotI; S, SacII.*

region known to contain the gene. For example, the *CFTR* (cystic fibrosis transmembrane regulator) gene was isolated in this way [5].

Initially, genetic linkage analyses suggested linkage of cystic fibrosis to markers on chromosome 7. Thereafter the CF chromosomal location was refined by genetic analysis. In addition to conventional linkage analysis which suggested a location at 7q31-q32, genetic clues regarding the location of the CF locus were obtained from the phenomenon of linkage disequilibrium, the non-random association of alleles at linked loci. In the case of diseases in which there is one major disease allele, such as cystic fibrosis and, most probably, Huntington's disease, the disease allele can be significantly associated with specific alleles at closely linked marker loci (*Figure 5.3*), whereas this phenomenon becomes increasingly less pronounced as the distance between disease and marker loci increases. Following the identification of closely linked flanking markers, the relevant chromosomal region was subjected to intensive molecular dissection involving the use of chromosome jumping and construction of contigs of overlapping genomic DNA fragments present in individual genomic clones. Candidate genes in this region were then identified by searching for highly conserved sequences or sequences which hybridized to mRNA (Section 4.3.2). The eventual identification of the *CFTR* gene was achieved by demonstrating a patient-specific mutation in one of the candidate genes; in the majority of CF patients, a deletion of three nucleotides was observed, resulting in the loss of a codon which specified phenylalanine at position 508 in a region of the protein which was expected to be important for gene expression.

5.2 Location and occurrence of pathological mutation

5.2.1 Intragenomic location of pathological mutation

Pathological mutation can occur at three types of DNA sequence:

(a) the DNA coding sequence of a gene;

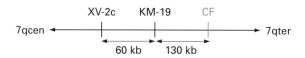

XV-2c/KM19 marker haplotype	XV-2c allele	KM-19 allele	Percentage of CF chromosomes with haplotype[a]	Percentage of normal chromosomes with haplotype[a]
A	1	1	7	30
B	1	2	86	14
C	2	1	3	44
D	2	2	4	12

Figure 5.3: *Linkage disequilibrium at the cystic fibrosis locus. [a]In the North American population.*

Table 5.2: *Location of pathological mutations in the Factor IX gene*

Location of mutation	Number of mutants	Unique molecular events
Promoter	12	8
Exons (a total of eight, spanning 1.4 kb)	352	179
Splice sites	24	19
Poly(A) site	0	0
Total	388	206

Data modified from [6] .

(b) intragenic non-coding sequences which are necessary for correct expression of the gene;

(c) regulatory sequences.

Many of the pathological mutations that have been described fall in the first category. For example, a database on world-wide studies of pathological mutations in the exons, promoter and exon/intron boundaries of the factor IX gene reveals 29 patients with partial or complete gene deletions and nearly 400 patients who have pathological intragenic point mutations or short intragenic insertions or deletions (<20 bp) [6]. In the latter group, only a small percentage of mutations were found at splice sites or in the promoter elements (*Table 5.2*). However, in other disorders abnormal splicing mutations may be common. In the case of the collagen disorder osteogenesis imperfecta, they constitute a very common pathological mutation which is second only in frequency to substitutions leading to replacement of the highly conserved, structurally important glycine residues. The collagen genes have small exons and a comparatively large number of introns (51), making them exceptional targets for splicing mutations. The latter are normally of the exon-skipping variety whereby point mutations or short deletions result in mutation of an acceptor splice site so that a whole exon is skipped in the splicing process (*Figure 5.4*).

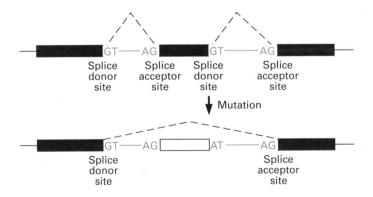

Figure 5.4: *Mutations at splice sites can cause exon skipping.*

In addition to promoter elements, other more distantly located regulatory elements may be sites of pathological mutation. For example, deletions which eliminate the β-globin LCR (see *Figure 1.7*) but leave the β-globin gene and its promoter intact result in almost complete abolition of β-globin gene expression and contribute to β-thalassemia. In the case of rare variants of α-thalassemia with mental retardation, the α-globin gene and its promoter may show no evidence of pathological mutation and the disease appears to map instead to the X chromosome, suggesting mutation in an X-linked gene which encodes a factor involved in regulating the expression of the α-globin gene.

Because of the large size of the human nuclear genome, most pathological mutation occurs in nuclear DNA sequences. However, due to the large amount of non-functional DNA in the nuclear genome most mutations in nuclear DNA are non-pathological. In comparison the mitochondrial genome is a small target for mutation (about 1/200 000 of the size of the nuclear genome). The proportion of clinical disease due to pathological mutation in the mitochondrial genome might therefore be expected to be extremely low. However, unlike the nuclear genome, the great bulk of the mitochondrial genome is composed of coding sequence and mutational rates in mitochondrial genes are about ten times higher than in their nuclear counterparts (possibly because of a reduced capacity for DNA repair). Accordingly, mutation in the mitochondrial genome is a significant contributor to human disease (see below).

5.2.2 Occurrence and frequency of pathological mutation

Fresh mutations can potentially occur at different stages of development. If a pathological mutation occurs in a single cell at a very early stage in embryonic development the mutation will ultimately be transmitted to a considerable proportion of the cells of the fully formed individual. An individual in whom a pathological mutation has arisen *de novo* is therefore a genetic mosaic, with a proportion of cells containing the mutation and the remainder lacking it. Somatic mosaicism refers to the mutational difference between different tissues. In the case of dominant disorders, the incidence and severity of the clinical phenotype will be dictated by the proportion of cells which carry the mutation in tissues in which the disease normally manifests. Although somatic mutations do not contribute to familial disease, they account for a significant proportion of human disease, notably the diverse types of cancer (see Section 5.4).

A mosaic in whom mosaicism extends to the gonads may also be a germ-line mosaic who will produce some gametes with the pathological mutation and some without, the proportion of each depending on the extent of mosaicism. Recent work on dominantly inherited osteogenesis imperfecta type II, a lethal disorder of collagen biosynthesis (Section 5.5.3), has identified eight cases of recurrence of the disorder in sibs and half sibs with the common parent being clinically asymptomatic or mildly affected (see [7] for an example). In all eight cases the common parent has been shown to be both a germ-line mosaic and a somatic mosaic. Consequently, the *de-novo* mutations in the common parent in each case have apparently occurred not in the maturation of the germ-line but very early on in the embryonic development of the mosaic parent. *De-novo* mutations like this which occur before the segregation of the germ-line and the soma may be a general phenomenon.

The incidence and range of pathological mutations can vary substantially between different disease loci. Dominant or X-linked recessive disorders in which the reproductive fitness of affected individuals is considerably diminished are necessarily characterized by a high frequency of *de-novo* mutation and generally by substantial mutational heterogeneity. In such cases constancy of disease gene frequency requires a large proportion of disease mutations to be fresh mutations in order to offset the large proportion of disease genes which are being removed from the population at each generation in non-reproducing patients. However, the new mutation rate in autosomal recessive disorders is relatively low; the frequency of asymptomatic carriers greatly exceeds that of affected individuals so that a relatively small proportion of disease mutations is lost in each generation.

Certain disease loci are thought to be susceptible to high frequencies of pathological mutation as a consequence of the structural organization of the underlying gene. For example, genes which are present on tandemly repeated DNA segments, or which contain intragenic tandem repeats, are particularly susceptible to potentially pathological intragenomic sequence exchanges in addition to random mutation (see next section). Additionally, it is probably highly significant that the underlying genes in single gene disorders such as Duchenne/Becker muscular dystrophy and NF1, which have very high mutation rates, are very large. Although the coding DNA component of the genes for dystrophin and NF1 presents a relatively modest target size for mutation, the large gene size admits a higher probability of mutation due to intragenic sequence exchanges such as recombination.

5.3 Genesis of pathological mutation

Although many mutations might be expected to arise as a result of essentially random errors of DNA replication or repair, characterized mutations in many genes show evidence of non-random mutations. In particular the CpG dinucleotide is known to be a hot spot for mutation, including pathological mutation, because of the inherent instability of methylated cytosine residues (Section 2.6). For example, the database of world-wide studies of hemophilia B reveals a high frequency of pathological mutation at CpG dinucleotides. In a group of nearly 400 patients with small-scale pathological mutations, nearly 50% of the mutations were found in multiple unrelated patients, and were due to mutation of a CpG dinucleotide in over 70% of cases (*Figure 5.5*). The same bias is found at a variety of loci; a recent survey of 152 pathological point mutations at 44 disease loci revealed that 32% were CpG → TpG or CpG → CpA transitions, a frequency 12-fold higher than expected [8]. In addition to CpG instability, other dinucleotide combinations which are prone to mutation are C/G-rich (GpG,GpC) while A/T-rich dinucleotides (TpA, ApA, TpT, ApT) are apparently more stable.

The organization of many individual genes also confers susceptibility to particular classes of potentially pathogenic sequence exchange mechanisms. In particular, short and long tandem repeats are mutational hot spots; genes which possess such sequences are normally characterized by frequent pathological deletion mutations. The outstanding example of a gene which is susceptible to pathogenic sequence exchange is the steroid 21-hydroxylase gene where 100% of recorded pathological mutations arise by such mechanisms (see Sections 5.3.1 and 5.3.3).

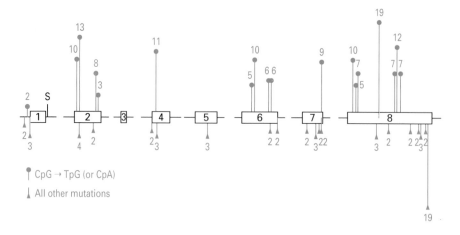

Figure 5.5: *Location of frequently occurring small pathological mutations (including insertions and deletions <20 bp) in the Factor IX gene. Boxes represent exons. Figures for mutations refer to the number of unrelated patients with an identical mutation.*

5.3.1 Disease due to large-scale DNA deletion and duplication

Disease due to large-scale DNA deletion includes rare cytogenetically visible deletions and smaller deletions which remove all or a significant portion of a gene. Large chromosomal deletions include interstitial deletions and terminal deletions. In the latter case the truncated human chromosome may be stabilized by the addition of telomeric repeats, as reported in the case of an α-thalassemia patient in whom the deletion breakpoint occurred 50 kb distal to the α-globin genes [9]. A similar mechanism may occur on other diseases which map to subterminal chromosomal regions and are associated with large deletions, including the Wolf–Hirschhorn (4p⁻) and Miller–Dieker (17p⁻) syndromes.

A number of genetic disorders are associated with a high frequency of gene inactivation due to deletion of large segments of DNA (*Table 5.3*). In the case of 21-hydroxylase deficiency, pathological deletions are consistently about 30 kb in length, approximately ten times the size of the *CYP21* gene which encodes 21-hydroxylase. The size of the deletions matches that of a single repeat unit in the tandemly duplicated *CYP21/C4* gene cluster (*Figure 2.3*), thereby suggesting the involvement of unequal cross-over or unequal sister chromatid exchange (Sections 2.7.2 and 2.7.4). Recently, the involvement of unequal cross-over in generating such large deletions has been strongly indicated by the demonstration of a *de-novo* deletion which contributes to 21-hydroxylase deficiency; the deletion has occurred on a chromosome which, because of exchange of flanking markers, is known to be a recombinant of homologous maternal chromosomes [10].

The deletions in 21-hydroxylase deficiency are uniform in size because the *CYP21/C4* gene organization represents an almost perfect large-scale VNTR system; deletions result from mispairing of large tandem repeats. In other genetic disorders large-scale deletions may be mediated by pairing of non-allelic interspersed repeats. For example, the *Alu* repeat occurs approximately once every 4 kb and mispairing

Table 5.3: *Human diseases associated with high frequencies of gene deletion*

Disease	Gene	Size of gene (kb)	Size of deletions (kb)	Approximate frequency of gene deletion in disease alleles (%)
X-linked ichthyosis (steroid sulfatase deficiency)	Steroid sulfatase	146	Often about 1900	90
Kearns–Sayre Syndrome	Mitochondrial genome	16.6	Heterogeneous 1.3–7.6	90
Duchenne muscular dystrophy	Dystrophin	2300	Heterogeneous	65
21-hydroxylase deficiency	CYP21 (21-hydroxylase)	3.3	Uniform; about 30	25

between such repeats has been found in a variety of disorders, notably familial hypercholesterolemia. The 45 kb LDL receptor gene has a relatively high density of *Alu* repeats (approximately one every 1.6 kb on the basis of the available sequence data). Out of eight pathological deletions in this gene in which the deletion endpoints were sequenced, seven were found to involve an *Alu* repeat, usually at both endpoints (*Table 5.4*).

Table 5.4: *Examples of pathological mutations in the LDL receptor gene resulting from* Alu-*mediated recombination*[a]

Pathological allele	Mutation class	Size (kb)	Location	Mutation mechanism
FH St Louis	Insertion	14.0	Intron 1–8	Duplication of exons 2 – 8 by *Alu–Alu* recombination
FH Rochester	Deletion	5.5	Intron 15 to exon 18	*Alu–Alu* recombination
FH Osaka-1	Deletion	7.8	Intron 15 to exon 18	*Alu–Alu* recombination
FH Osaka-2	Deletion	12.0	Intron 6–14	*Alu–Alu* recombination
FH London-1	Deletion	4.0	Intron 12–14	*Alu–Alu* recombination
FH Potenza	Deletion	5.0	Exon 13 to intron 15	Exon–*Alu* recombination

[a]Data abstracted from reference 11.
FH, familial hypercholesterolemia.

Other interspersed repeats may also be involved in mediating deletions by non-homologous recombination. For example, although there are 48 *Alu* sequences within 66.5 kb of DNA in the growth hormone gene cluster, hot spots for growth hormone gene deletions occur outside the *Alu* repeats, often involving two 594 bp repeat elements which flank the *GH1* gene and are about 99% homologous in sequence [12]. In some cases the interspersed repeats may be distantly located. X-linked ichthyosis is frequently due to large deletions at the steroid sulfatase locus; the 146 kb steroid sulfatase gene is deleted in about 90% of patients, with the deletions usually spanning about 1.9 Mb. Such giant deletions usually occur at breakpoints within different repeats belonging to a low copy number repetitive DNA family, DXS278 [13]. Part of the individual repeats is composed of VNTR sequences and it has been proposed that folding of the X chromosome in the nucleus may result in close physical proximity of the repeats.

In several cases, the endpoints of deletions involve very short direct repeats. For example, the breakpoints in numerous pathological deletions of the mitochondrial genome occur at perfect or almost perfect short direct repeats. Of these, the most common is a deletion of 4977 bp which has been found in multiple patients with Kearns-Sayre Syndrome, an encephalomyopathy characterized by external ophthalmoplegia, ptosis, ataxia and cataract. The deletion results in elimination of the intervening sequence between two perfect 13 bp repeats and loss of the sequence of one of the repeats (*Figure 5.6*). Such mutations have been postulated to arise by a slip-replication mechanism [14].

DNA sequencing of the deletion breakpoints in the dystrophin gene of different muscular dystrophy patients has revealed a common 6 bp motif, but the mechanism by which such deletions occur is not known at present. In addition to large deletions, large-scale DNA duplications are found in the dystrophin gene of about 5% of patients. This may suggest the involvement of unequal cross-over or unequal sister chromatid exchange. Possibly, many such intragenic duplications are non-pathological and go

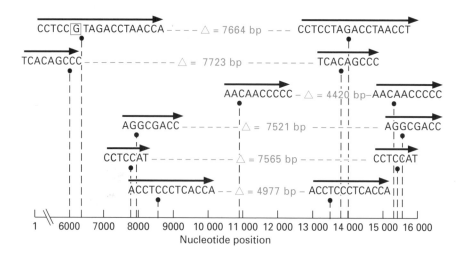

Figure 5.6: *Short direct repeats mark the endpoints of pathological deletions in the mitochondrial genome.* △ *– deletion.*

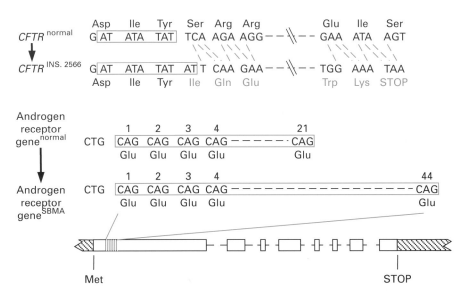

Figure 5.7: *Pathological mutation due to variation in the number of short tandem repeats.*

undetected in clinically asymptomatic individuals. Other examples of large-scale pathological DNA duplications are rare, but include duplications of 8 kb of the mitochondrial genome in patients with mitochondrial myopathy.

5.3.2 Disease due to variation in copy number of short intragenic tandem repeats

A number of pathological mutations have been shown to involve deletion or duplication of short DNA segments occurring in tandemly duplicated repeats. Depending on the size of the repeat unit, the alteration may result in frameshift mutations which are often pathological as a result of the introduction of a premature termination codon in the new reading frame (*Figure 5.7*). Variation in tandem trinucleotide and hexanucleotide repeats would not introduce a frameshift mutation. However, large-scale tandem repetition of trinucleotide repeat units has been implicated in DNA instability which may be pathogenic. For example, the first exon of the androgen receptor gene has a stretch of about 20 tandemly repeated CAG codons which specify a polyglutamine tract (*Figure 5.7*). Amplification of this repeat (possibly by slippage replication, or possibly even by unequal cross-over or unequal sister chromatid recombination) appears to be specifically associated with X-linked spinal and bulbar muscular atrophy and may even be the cause of the disease [15].

A close parallel to the above observation has very recently been reported in the *FMR-1* gene, the candidate gene for fragile X-linked mental retardation. Near the 5' end of this gene there are about 40 tandemly repeated CGG units, encoding a polyarginine segment. The length of the poly CGG tract appears to vary considerably in normal chromosomes and has been considered to be the location of the pathological instability that marks this chromosomal region [16].

5.3.3 Disease due to gene conversion-like events

As with unequal cross-over, the strongest evidence for the occurrence of gene conversion-like events in the human genome has been obtained by analysis of the molecular pathology of 21-hydroxylase deficiency. All of the approximately 75% of gene lesions which are not due to gene deletion show evidence of replacement of all, or a small portion of the normally functional *CYP21* gene by a copy of the analogous sequence from the nearby *CYP21P* pseudogene. The large-scale replacement of *CYP21* by a copy of *CYP21P* could possibly occur by multiple independent recombination events, including at least one unequal cross-over, or unequal sister chromatid exchange. However, the great majority of the *CYP21* → *CYP21P* replacement events involve only a small portion of the CYP21 genes, suggestive of a micro-gene conversion event which introduces a pathological point mutation into the *CYP21* gene copied from the *CYP21P* pseudogene (*Table 5.5*). In the only recorded example of a *de-novo* pathological point mutation in the *CYP21* gene, the proposed micro-conversion event introduces a maximum of 390 bp of defective DNA sequence copied from the *CYP21P* pseudogene [17].

Table 5.5: *Pathological point mutations in the CYP21 (21-hydroxylase) gene*

Location of mutation	Normal CYP21 gene sequence	Mutant CYP21 gene sequence	CYP21P gene sequence (pseudogene)
Intron 2	cccacctcc	cccaGctcc	cccaGctcc
Exon 3 (codons 110–112)	gga gac tac tc Gly Asp Tyr Ser	g(.. ... ...)tc V al	g(.. ... ...)tc
Exon 4 (codon 172)	atc atc tgt Ile Ile Cys	atc aAc tgt Ile Asn Cys	atc aAc tgt
Exon 6 (codons 235–238)	atc gtg gag atg Ile Val Glu Met	aAc gAg gag aAg Asn Glu Glu Lys	aAc gAg gag aAg
Exon 7 (codon 281)	cac gtg cac His Val His	cac Ttg cac His Leu His	cac Ttg cac
Exon 8 (codon 318)	ctg cag gag Leu Gln Glu	ctg Tag gag Leu STOP	ctg Tag gag
Exon 8 (codon 356)	ctg cgg ccc Leu Arg Pro	ctg Tgg ccc Leu Trp Pro	ctg Tgg ccc

Evidence for gene conversion contributing to pathogenesis in other systems is meager. Mutations in the human β-glucocerebrosidase gene causing Gaucher's disease occasionally show evidence suggestive of a gene conversion from the neighboring β-glucocerebrosidase pseudogene. The steroid 11-hydroxylase gene cluster is another gene–pseudogene system which is a likely candidate for pathogenesis due to gene conversion.

5.3.4 Disease due to translocation

Rare chromosomal translocation events have been identified as contributing to the pathology of a number of human diseases. In most cases, this is because a breakpoint within the gene disrupts gene expression. Occasionally, however, a translocation can result in unwanted activation of expression of an oncogene, a gene involved in regulation of cell growth and division, with the resultant development of a tumor (Section 5.4.2). Very little is known about the factors which govern the breakpoints of translocations as they have not been characterized in the vast majority of cases. In some cases, apparent non-homologous recombination between different chromosomes may be facilitated by interaction between homologous sequences at the recombination breakpoints.

5.3.5 Disease due to DNA transposition

Defective gene expression due to DNA transposition is comparatively rare and represents only a small component of molecular pathology. However, examples have been recorded of genetic deficiency due to insertional inactivation by transposons. For example, in one study hemophilia A was found to arise in two out of 140 unrelated patients as a result of a *de-novo* insertion of a *Kpn (L1)* repeat into an exon of the Factor VIII gene [18]. Occasional cases of hemophilia B and of NF1 have also been shown to be due to transposition mutations involving insertion of *Alu* repeats into the Factor

Table 5.6: *Examples of breakpoints observed in specific tumors*

Breakpoint	Tumors	Comments
1p36	ML, AML, neuroblastoma	
1q11–q12	MEL	
1q21	AML, AC of bladder, uterus and breast	
2p23	AML, ML	
3p21–p13	AC of lung, kidney, breast and ovary	
3q21, 3q26	AML, MDS, MPS	Inv(3)(q21q26); t(3:3)(q21;q26)
9q34	AML, MPD, CML, ALL	*ABL* oncogene site
10q23–q24	T-ALL, ML, AC of prostate	*ETS1* oncogene site
11p13	Wilms tumor	Site of Wilms tumor gene
13q14	Retinoblastoma	Site of retinoblastoma gene
14q11	T-cell malignant lymphomas	Site of a T-cell receptor gene cluster
14q32	B-cell malignant lymphomas	Site of heavy chain immunoglobulin gene cluster
22q11–q13	ML, BL, ALL, CML, AML, MN	Site of light chain immunoglobulin gene cluster and a meningioma suppressor gene

AC, adenocarcinoma; ALL, acute lymphoblastoid leukemia; AML, acute myeloid leukemia; BL, Burkitt's lymphoma; CML, chronic myeloid leukemia; MDS, myelodysplastic syndrome; MEL, malignant melanoma; ML, malignant lymphomas; MN, meningioma; MPD, myeloproliferative disorder.

IX and NF1 genes respectively. Additionally, a number of other examples have been recorded of pathogenesis due to intragenic insertion of undefined DNA sequences.

5.4 Neoplasia

A variety of aneuploid states, single gene disorders and polygenic conditions predispose to human cancer. At the chromosomal level, a large number of consistently occurring structural chromosome changes have been reported to be associated with specific human malignancies. The karyotypes of cancer cells are frequently anomalous and include many non-specific abnormalities, in addition to tumor-specific changes. By mid-1990, a total of 179 non-random structural chromosomal changes were identified in 51 different types of neoplastic disorders, including hematological diseases, malignant lymphomas and solid tumors (*Table 5.6*). Recently, molecular analyses have led to the mapping and isolation of a large number of genes which are directly implicated in human cancer. The genes can be considered to fall into three classes: DNA repair genes, oncogenes and tumor suppressor genes.

5.4.1 Deficiency of DNA repair genes

A variety of cellular mechanisms exist for repairing or tolerating damage in genomic DNA. Mutations in genes which encode proteins involved in DNA repair result in predisposition to malignancy. At the cellular level they are often characterized by a high frequency of spontaneous chromosome aberrations and by hypersensitivity to u.v. light and/or ionizing radiation. The disorders ataxia telangectasia and xeroderma pigmentosum show an autosomal recessive mode of inheritance and considerable genetic heterogeneity (see *Table 5.7*).

5.4.2 Cellular oncogenes

Oncogenes are genes which can transform a normal cell to a tumor cell. Initially such genes were identified and characterized in viruses which can induce neoplastic transformation. More recently, several different cellular counterparts of viral oncogenes have been described in different species including humans, where their normal function, as proto-oncogenes, is to control cell growth. Many of the estimated 60–70 human proto-oncogenes have been characterized at the molecular level and four major classes can be distinguished (*Table 5.8*). Some are *G-proteins*, proteins which can bind GTP and which have intrinsic GTPase activity, permitting a role as second messengers in the cell. In addition, several oncogene products, oncoproteins, act as transcriptional factors. The precise mode of action of many oncogenes remains to be elucidated, although some are known to encode tyrosine kinases which, by phosphorylating tyrosine residues on their target proteins, can modulate the activity of the target proteins.

Cellular proto-oncogenes are thought to become oncogenic in response to mutations which cause altered, enhanced or inappropriate constitutive expression. Most mutations occur at the somatic level so that the resulting identified cancers appear as sporadic cases. Identified point mutations include mutations which abolish the inherent GTPase activity of transducer-encoding genes and an exceptional single

Table 5.7: *Examples of inherited single gene cancer disorders*

Disorder	Common tumor types	Type of defect[a]	Gene and location
Ataxia telangiectasia	Lymphoma	DNA repair	Many but unidentified, e.g. 11q22–q23
Xerodermal pigmentosum	Skin carcinoma	DNA repair	*ERCC3,* 2q21; *ERCC5,* 13q22-q34, others on 1q, 9, 15, etc.
Familial adenomatous polyposis	Adenocarcinoma of the colon	TS	*APC,* 5q21
Early onset hereditary breast cancer	Breast carcinoma	TS	Unidentified, 17q21
Li–Fraumeni Syndrome	Breast cancer and other neoplasms	TS	*TP53,* 17p13.1
Multiple endocrine neoplasia			
1	Pituitary and pancreatic adenoma	TS	Unidentified, 11q13
2A	Pheochromocytoma, medullary thyroid carcinoma, parathyroid adenoma	TS	Unidentified, 10
2B	Pheochromocytoma, medullary thyroid carcinoma	TS	Unidentified, 10
NF1	Fibrosarcoma, optic glioma	TS	*NF1,* 17q11.2
NF2	Acoustic neuroma, schwannoma, meningioma	TS	Unidentified, 22q11-q13.1
Retinoblastoma	Retinal embryonic tumor, osteosarcoma	TS	*RB1,* 13q14.2
Tuberose sclerosis	Cardiac rhabdomyoma, renal angiomyolipoma	TS	Unidentified, 9q, 11q14–q23
von Hippel–Lindau	CNS hemangioblastoma renal carcinoma	TS	Unidentified, 3p25–p26
Wilms tumor	Renal embryonic tumor	TS	*WT1,* 11p13 and others

[a]TS, tumor suppressor gene.
CNS, central nervous system.

nucleotide substitution within an intron of the *HRAS* (Harvey ras) gene which results in a tenfold increase in gene expression [19].

Proto-oncogenes can also be activated by mutations which do not alter their coding sequence, including extragenic insertions, gene amplification and chromosomal translocations. The latter often occur in B or T lymphocytes and appear to result from

Table 5.8: *Examples of human oncogenes*

Oncogene	Chromosomal location	Function of product	Site of action
CSF1R (*FMS*)	5q33–q35	Receptor for CSF-1 (macrophage colony stimulating factor)	Plasma membrane
EGFR (*ERBB*)	7p12–p13	Receptor for epidermal growth factor	Plasma membrane
ETS1	11q23.3	Transcription factors	Nucleus
ETS2	21q22.3	– bind to PEA 3 motif	
FOS	14q24.3	Transcription factors	Nucleus
JUN	1p31–p32	– bind to AP-1 motif	
HRAS	11p15.5	G protein	Cytoplasm
KRAS2	12p12.1	G protein	Cytoplasm
NRAS	1p13	G protein	Cytoplasm
PDGFB (*SIS*)	22q12–q13	Platelet-derived growth factor β chain	Secreted

malfunction of the programmed recombination events which normally generate functional immunoglobulin or T-cell receptor genes, respectively – see Section 2.7.3. The recombinase involved in these processes may occasionally recognize sequences which are coincidentally related to its normal recognition sequence. If such sequences are located at a proto-oncogene locus, chromosomal translocation may occur involv-

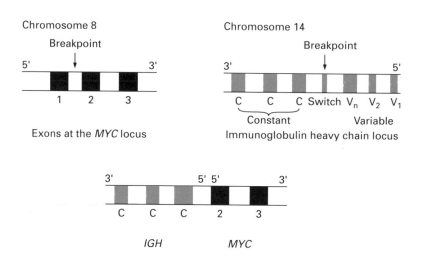

Figure 5.8: *Molecular basis of the translocation between the* MYC *locus on chromosome 8 and the immunoglobulin heavy chain locus on chromosome 14 which occurs in the majority of cases of Burkitt's lymphoma.*

ing recombination between the proto-oncogene and an immunoglobulin gene or T-cell receptor gene (*Figure 5.8*). Some translocations are known to produce truncated or fusion genes with aberrant responses to control sequences. Other translocations remove individual oncogenes to a different chromosomal environment where they are free from the effects of the control sequences which normally regulate them. Many of the cytogenetically identified tumor-specific breakpoints are known to lie near oncogenes.

5.4.3 Tumor suppressor genes

Most inherited cancers involve genetic defects in a class of genes which have been called tumor suppressor genes (or anti-oncogenes), defects in which represent the most common cause of inherited cancers (*Table 5.7*). Such genes are known to contribute to dominantly inherited cancers by a mechanism in which tumorigenesis is recessive at the cellular level (i.e. if the normal gene product is added to the tumor cell, the normal cell phenotype is restored). In such cases tumorigenesis is thought to occur as a consequence of inactivation of both gene copies of an autosomal tumor suppressor gene. Normally, tumor suppressor genes express a product which can suppress the

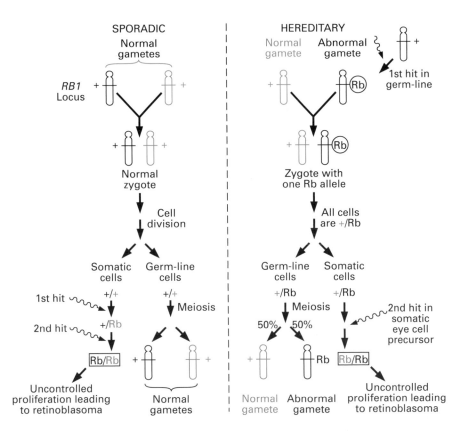

Figure 5.9: *Origin of hereditary and sporadic retinoblastoma.*

expression or function of other genes concerned with growth and proliferation of particular cell types. Inactivation of one of the two copies of an autosomal tumor suppressor gene results in 50% of the normal tumor suppressor product and is normally accompanied by minor or no apparent alteration of the normal cell phenotype. However, homozygous gene inactivation at a tumor suppressor locus leads to absence of a tumor suppressor product and can therefore lead to tumorigenesis.

Disorders of tumor suppressor genes may occur in both hereditary and non-hereditary forms. Study of some such disorders, notably the eye cancer, retinoblastoma, has revealed that the mechanism underlying inactivation of a tumor suppressor gene is consistent with a 'two-hit' mutation model [20]. The first mutational hit is often a subtle mutation which may occur either in the germ-line or in a somatic cell precursor of the type of cell in which the tumor develops (*Figure 5.9*). The second mutational hit occurs subsequently in the somatic cell from which the tumor arises and is often a large-scale mutation (*Figure 5.10*). Such mutations are often detectable cytogenetically, or by comparison of tumor and blood samples from the same individual to establish the loss in the tumor of constitutional heterozygosity (Section 4.2.6). Individuals in whom both mutations occur somatically show no family history of the disease and do not transmit it to subsequent generations. If the first mutation occurs in the germ-line, 50% of the descendants will have one inactivated allele and a dominant pattern of inheritance is established. Often such individuals may present with bilateral tumors because of secondary mutational hits in different cells. Although it is known that the products of oncogenes and tumor suppressor genes can interact, the mode of action of most tumor suppressors is presently unknown. One candidate tumor suppressor gene, *PTPG*, which maps to 3p21, encodes a receptor protein tyrosine phosphatase which may reverse the effects of protein tyrosine kinases, many of which are oncogenes [21]. The *TP53* gene, defects in which are the most common known cause of human cancer (see Section 6.1.2), encodes a sequence-specific DNA binding protein, p53, whose function may be mediated by its ability to bind to specific DNA sites in the human genome. Mutations which occur in the *TP53* gene not only cause loss of p53 tumor suppressor function but can also activate p53 as an oncogene. Activated p53 mutants

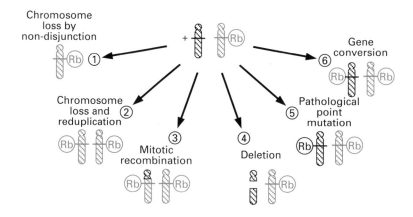

Figure 5.10: *Mechanisms whereby a recessive mutation at the retinoblastoma locus might be unmasked.*

are known to cause tumor progression by a remarkable dominant negative effect; co-translation of activated mutant p53 with wild-type p53 drives the wild-type allele into the mutant conformation [22].

5.5 Expression of pathological mutations

The contribution of pathological mutation in a single gene to the clinical phenotype depends on:

(a) the effect of the mutations on the expression of the relevant gene;

(b) dominance or recessivity of the mutant gene in relation to normal gene copies;

Table 5.9: *Effect of mutation on gene function*

Location and nature of mutation	Effect on gene function	Comments
Extragenic mutation	Normally none	Rare mutations may result in inactivation of distant regulatory elements required for normal gene expression (see Section 5.2.1)
Multigene deletion	Abolition	Associated with contiguous gene syndromes
Whole gene deletion	Abolition	
Whole exon deletion	Abolition or modification	May cause shift in reading frame; protein often unstable
Within exon	Abolition	If loss/change of key amino acids, shift of the reading frame or introduction of premature stop codon.
	Modification	If non-conservative substitutions, small in-frame insertions, or other mutations at some locations
	None	Conservative/silent substitutions or mutation at non-essential sites
Whole intron deletion	None	
Splice-site mutation	Abolition or modulation of expression	Conserved GT and AG signals are critically important for normal gene expression
Promoter mutation	Abolition or modulation of expression	Deletion, insertion or substitution of nucleotides within promoter may alter expression. Complete deletion abolishes function
Mutation of termination codon	Modification	Additional amino acids are included at the end of the protein until another stop codon is reached.
Mutation of poly(A) signal	Abolition or modulation of expression	Deletion, insertion or substitution of nucleotides within poly(A) site may alter expression. Complete deletion abolishes function.
Elsewhere in introns/UTS	Usually none	

(c) the proportion and nature of cells in which the mutant gene is present;

(d) in some cases, the parental origin of the mutation.

5.5.1 Effect of mutation on individual gene expression

Recent molecular advances have established much information concerning the nature of the molecular defects underlying certain single gene disorders and several different classes of mutation have been described (*Table 5.9*). Essentially, such mutations can abolish or reduce normal gene expression or result in inappropriate or aberrant gene expression. In general, there is a correlation between the degree to which a pathological mutation affects normal gene expression and the severity of the resultant clinical phenotype. When the pathological effect is due to deficiency of a gene product, gene deletions or other mutations which lead to absence or complete inactivation of the normal gene product (e.g. nonsense mutations or frameshift mutations which introduce a premature termination codon) are normally associated with severe clinical phenotypes. However, in some disorders of multimeric proteins, such mutations can produce much milder phenotypes than more subtle mutations (see Section 5.5.3).

Small intragenic deletions may produce variable phenotypes depending on whether they introduce an extensive shift in the reading frame. For example, the severely

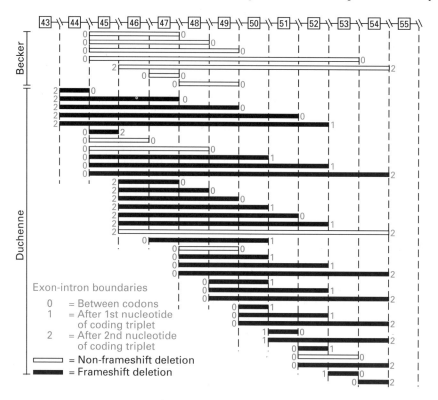

Figure 5.11: *Deletions in the central part of the dystrophin gene associated with Becker and Duchenne muscular dystrophy. Numbered boxes represent exons 43–55.*

crippling Duchenne muscular dystrophy and the much milder Becker's muscular dystrophy are now known to be due to mutation at the same gene locus, dystrophin. Both disorders are mutationally heterogeneous, but are often characterized by intragenic deletions. The difference in severity between the clinical phenotypes has recently been ascribed to the effect of the deletion on altering the translational reading frame. Deletions which do not alter the reading frame result in elimination of coding sequence within the deleted region but do not affect the downstream coding sequence, and are normally associated with the relatively mild Becker's muscular dystrophy [23]. Deletions that alter the reading frame have a more marked effect because they alter the coding sequence downstream of the deleted region and are more character-istic of Duchenne muscular dystrophy (*Figure 5.11*). In some cases a mild clinical phenotype may result from very large intragenic deletions, including one exceptional 700 kb deletion resulting in the removal of 46% of the dystrophin coding sequence [24].

5.5.2 Recessive disease alleles

Recessive disease alleles are normally non-functional or produce a limited amount of normal gene product. Heterozygotes with one recessive disease allele and one normal allele can therefore produce 50% or more of the normal product and for many gene products (e.g. enzymes) this deficiency is not sufficient to produce clinical symptoms (i.e. the disease allele is recessive to the normal allele). If the disease locus is autosomal, the clinical phenotype is determined by the compound effect of pathologi-cal mutations at two alleles. In individuals where the two disease alleles are different (compound heterozygotes), the phenotype is dictated by the disease allele which has the lesser effect on gene expression. Because boys normally have a single X chromosome (and are therefore hemizygous for all genes on the X chromosome), a single mutant allele can result in X-linked disease in the absence of a normal allele.

Recessive disease alleles often occur in genes which produce enzymes, but may also occur in other types of protein such as carrier proteins (e.g. α- and β-globin), or regulatory genes (e.g. tumor suppressor genes). In the latter case, although the mutant allele is recessive to the normal allele in its phenotypic expression, the disease is actually transmitted as a dominantly inherited disorder (see Section 5.4.3).

5.5.3 Dominant disease alleles

Dominantly inherited disorders are characterized by the expression of clinical symptoms in the heterozygote so that the disease allele is dominant over the normal allele. Dominant pathological mutations often result in deficiency or abnormal gene expression. Additionally, many examples are known of dominant oncogenic muta-tions which are due to constitutive expression or over-expression of a normal gene product. Generally, dominant pathological alleles occur in a gene which normally produces a non-enzymatic protein, such as a receptor, carrier protein, structural protein or regulatory protein. For example, familial hypercholesterolemia can result from a single mutation in the LDL receptor gene resulting in deficiency of functional LDL receptors. Because LDL regulates HMG-CoA reductase, the rate-limiting enzyme in endogenous cholesterol biosynthesis, heterozygotes with 50% functional

LDL receptors demonstrate normal regulation by LDL of HMG-CoA reductase but only at the expense of an increased LDL cholesterol level and increased susceptibility to premature atherosclerotic heart disease. Homozygotes (who usually are really compound heterozygotes) have little or no functional LDL receptor, very high levels of cholesterol and usually develop heart disease in their twenties.

Many of the inherited disorders of the structural protein collagen are also characterized by dominant disease alleles. Collagens are synthesized as a triple helix comprising three long polypeptide chains. In individual molecules the three chains may be synthesized by a single type of gene or by two different genes. For example, type I collagen which is predominantly found in bone, tendon and skin, consists of two $\alpha 1$(I) chains (encoded by the *COL1A1* gene on 17q) and an $\alpha 2$(I) chain (encoded by the *COL1A2* gene on 7q). Unexpectedly, mutations which completely inactivate an α chain gene can give rise to a comparatively mild clinical phenotype, type I osteogenesis imperfecta, whereas more subtle pathological mutations in the same gene, for example, point mutations and rearrangements, result in the much more severe type II osteogenesis imperfecta. This apparent paradox has recently been resolved. Complete inactivation of one allele of the *COL1A1* gene results in 50% production of the normal $\alpha 1$(I) chains, and 50% reduction in the amount of procollagen molecules, with the excess $\alpha 2$(I) chains being degraded (see *Figure 5.12*). However, subtle mutations resulting in the production of near normal amounts of a mutant *COL1A1* product will result in abnormalities in 50% of the $\alpha 1$(I) product and in 75% of the procollagen molecules.

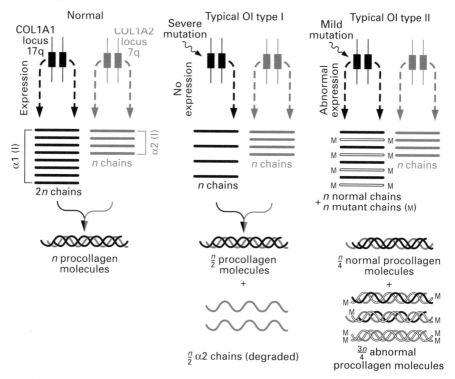

Figure 5.12: *Defective collagen synthesis in osteogenesis imperfecta (OI) types I and II.*

Table 5.10: *Mitochondrial disease*

Type of mutation	Disease	Clinical phenotype	Inheritance of disease
Point mutation at position 11778 (Arg→His in mitochondrial *ND4* gene) or at position 3460 (Ala→Thr) in mitochondrial *ND1* gene	Leber's hereditary optic neuropathy	Optic atrophy plus other neurological symptoms	Maternal in familial cases
Point mutation at position 8993 (Leu→Arg) in sub-unit 6 of mitochondrial ATPase	Variable neurological syndrome	Retinitis pigmentosa plus ataxia, seizures, dementia, proximal muscle weakness	Sporadic
Point mutation in TψC loop of mitochondrial tRNA^{Lys} gene	MERFF	Myoclonic epilepsy with ragged red fibers (MERRF) in skeletal muscle	Maternal in familial cases
Large deletions of mitochondrial genome	(1) Kearns–Sayre Syndrome	Progressive external ophthalmoplegia (PEO) plus other symptoms	Maternal in familial cases
	(2) Other PEO conditions		Maternal or autosomal dominant
Large (≈ 8 kb) tandem duplication in mitochondrial genome	Kearns–Sayre Syndrome	As above	Sporadic
Severe depletion of mitochondrial DNA	PEO plus other symptoms	Variable tissue expression, can affect muscle or liver in siblings	Defect in nuclear gene?

5.5.4 Mitochondrial disease

Normal mitochondrial function is unusual in being dependent on the complementation of two genomes: the nuclear genome in which genes are transmitted as allelic Mendelian traits, and the mitochondrial genome which is maternally inherited. In principle, therefore, mitochondrial disease could result from mutations in either nuclear or mitochondrial genes which specify mitochondrial products, or from mutations in nuclear genes which control mitochondrial DNA expression. However, most mitochondrial disorders that have been characterized, including certain neurodegenerative and neuromuscular diseases, are due to mutation in the mitochondrial genome [25] (*Table 5.10*), although at least one example is known of mitochondrial dysfunction due to nuclear mutation [26].

In the case of disease due to mutation in the mitochondrial genome, the severity of the clinical phenotype is normally proportional to the number of mutant mitochondrial genome copies. The consequences of *de-novo* mutation occurring in the mitochondrial genome are complicated by the existence of thousands of mitochondrial genomes per cell and the random segregation of mitochondria into daughter cells during cell division. In cells with a mixture of mutant and wild-type mitochondrial DNAs (heteroplasmy), the mitochondrial genotype can alter following successive cell divisions so that some cell lineages can drift towards pure mutant mitochondrial DNA (homoplasmy), others towards pure wild-type, while others remain heteroplasmic. It is noticeable that mitochondrial mutations are often large deletions. The resulting small mutant genome may be at a selective advantage in terms of replication efficiency, so that the mutant genome may preferentially accumulate.

5.5.5 Mosaicism

The expression in a genetic mosaic (Section 5.2.2) of a pathological allele in some cells, but not in others, can result in variation in the severity of the clinical phenotype in proportion to the amount of relevant cells expressing the disease allele. All women are mosaics because in some of their cells the paternal X chromosome is inactivated (Section 2.2) while in others the maternal X chromosome is inactivated. A small percentage of obligate female carriers of X-linked disease do express disease symptoms. It is possible that such apparent manifesting heterozygotes may include genuine heterozygotes in whom, by chance, the X chromosome which carries the normal allele is inactivated in most of the relevant cells. In many cases, however, such individuals may be undetected compound heterozygotes, or individuals in which there has been a secondary mutation at the X-chromosome inactivation center, as suggested in the reported case of female to female transmission of X-linked hemophilia B. Germ-line mosaicism has been recorded in a number of different disorders and can raise problems for prenatal diagnosis.

5.5.6 Genomic imprinting

Evidence for differential function of maternally and paternally derived chromosomes and alleles has accumulated from the study of several mammalian species, notably the mouse, but increasingly also from human studies [27]. For example, a conceptus with a 46,XX karyotype which is exclusively paternal in origin always develops into a hydatidiform mole consisting of abnormal chorionic trophoblast with no embryo.

Growth abnormalities are known to result from uniparental disomy: the inheritance of both sets of a specific chromosome from a single parent instead of one from each parent. This may occur following the production of a zygote which has three copies of a specific chromosome (trisomy) or only one copy (monosomy). During early development, selection against cells with such aberrant karyotypes may lead to the selective growth of cells which have regained the normal 46,XX or 46,XY karyotype by loss or duplication of the relevant chromosome. Assuming equal probabilities of loss of maternally and paternally inherited homologs, there is a one in three chance that chromosome loss following initial trisomy will develop into uniparental disomy. However, chromosome duplication following initial monosomy will always lead to uniparental disomy.

In mice, the paternally derived X chromosome is preferentially inactivated in placental tissue, consistent with some mechanism for labeling and distinguishing

Table 5.11: *Examples of human disease in which the involvement of genomic imprinting has been suspected*

Disease	Nature of differential expression	Chromosomal location
Angelman Syndrome	*De-novo* deletions of 15q11–q13 on maternally inherited chromosomes; non-deletion cases occasionally show paternal uniparental disomy	15q11–q13
Prader–Willi Syndrome	*De-novo* deletions of 15q11–q13 on paternally inherited chromosomes; other cases include examples of maternal uniparental disomy	15q11–q13
Beckwith–Wiedemann Syndrome	Some cases with trisomy for 11p15.5, and paternal origin of the duplicated segment	11p15.5
Huntington's disease	Earlier onset with paternal transmission	4p16
Cerebellar ataxia	Earlier onset with paternal transmission	?
Spinocerebellar ataxia	Earlier onset with paternal transmission	6p24–p21
Myotonic dystrophy	Congenital form almost exclusively in children born to affected mothers	19q13
NF1	Increased severity with maternal transmission	17q11.2
NF2	Earlier onset with maternal transmission	22q11–q13.1
Wilm's tumor	Loss of maternal alleles in sporadic tumors; 7/8 *de-novo* germ-line deletions involving 11p13 are of paternal origin	15
Osteosarcoma	Loss of maternal alleles in sporadic tumors	13q
Cystic fibrosis	Rare cases with accompanying growth retardation and uniparental disomy	7

paternally and maternally derived homologs. Consequently, it has been hypothesized that during the production of sperm cells and egg cells, chromosomes are labeled, probably by DNA methylation, in such a way as to record their parental origin (genomic imprinting).

The significance of genomic imprinting in relation to human disease is that it affords the possibility of differential expression of the same inherited mutation depending on the sex of the transmitting parent. Deletions in the q11–13 region of a paternally derived chromosome 15 are commonly associated with Prader–Willi Syndrome whereas deletions in the same apparent chromosomal region of a maternally derived chromosome 15 are found in Angelman Syndrome patients. However, although non-deletion cases of Prader–Willi Syndrome often show maternal chromosome 15 disomy, paternal chromosome 15 isodisomy cases are relatively rare in Angelman's Syndrome. Differences in the expression of paternally and maternally inherited mutations have been documented at a variety of other disease loci and in some cases may be due to genome imprinting (*Table 5.11*).

References

1. Dryja, T.P., McGee, T.L., Reichel, E. *et al.* (1990) *Nature,* **343**, 364.
2. Dietz, H.C., Cutting, G.R., Pyeritz, R.E. *et al.* (1991) *Nature,* **352**, 337.
3. Monaco, T. and Kunkel, L.M. (1987) *Trends Genet.* **3**, 35.
4. Fountain, J.W., Wallace M.R., Bruce, M.A. *et al.* (1989) *Science,* **244**, 1085.
5. Rommens, J.M., Iannuzi, M.C., Kerem, B.-S. *et al.* (1989) *Science,* **245**, 1059.
6. Gianelli, F., Green, P.M., High, K.A. *et al.* (1991) *Nucl. Acid Res.,* **19**, 2193.
7. Cohn, D.H., Starman, B.J., Blumberg, B. and Byers, P.H. (1990) *Am. J. Hum. Genet.,* **46**, 591.
8. Cooper, D.N. and Krawczak, M. (1990) *Hum. Genet.,* **85**, 55.
9. Wilkie, A.O.M., Lamb, J., Harris, P.C., Finney, R.D. and Higgs, D.R. (1990) *Nature,* **346**, 868.
10. Sinnott, P.J., Collier, S., Costigan, C., Dyer, P.A., Harris, R.and Strachan, T. (1990). *Proc. Natl Acad. Sci. USA,* **87**, 2107.
11. Hobbs, H.H., Russell, D.W., Brown, M.S. and Golding, J.L. (1990) *Ann. Rev. Genet.,* **24**, 133.
12. Vnencak-Jones, C.L., Phillips II, J.A., Chen, E.Y. and Seeburg, P.H. (1988) *Proc. Natl Acad. Sci. USA,* **85**, 5615.
13. Yen, P.H., Li, X.-M., Tsai, S.P., Johnson, C., Mohandas, T. and Shapiro, L.J. (1990) *Cell,* **61**, 603.
14. Shoffner, J.M., Lott, M.T., Voljavec, A.S., Soueidan, S.A., Costigan, D.A. and Wallace, D.C. (1989) *Proc. Natl Acad. Sci. USA,* **86**, 7952.
15. La Spada, A.R., Wilson, E.M., Lubahn, D.B., Harding, A.E. and Fischbeck, K.H. (1991) *Nature,* **352**, 77.
16. Verkerk, A.J.M.H., Pieretti, M., Sutcliffe, J.S. *et al.* (1991) *Cell,* **65**, 905.
17. Collier, S., Sinnott, P.J., Harris, R. and Strachan, T. *J. Med. Genet.,* **28**, 562.
18. Kazazian, H.H., Wong, C., Youssoufian, H. *et al.* (1988) *Nature,* **332**, 164.
19. Cohen, J.B. and Levinson, A.D. (1988) *Nature,* **334**, 119.
20. Knudson, A.G., Jr. (1986) *Ann. Rev. Genet.,* **20**, 231.
21. LaForgia, S., Morse, B., Levy, J. *et al.* (1991) *Proc. Natl Acad. Sci. USA,* **88**, 5036.
22. Milner, J. and Medcalf, E.A. (1991) *Cell,* **65**, 765.
23. Koenig, M., Beggs, A.H., Moyer, M. *et al.* (1989) *Am. J. Hum. Genet.,* **45**, 498.
24. England, S.B., Nicholson, L.V.B., Johnson, M.A. *et al.* (1990) *Nature,* **343**, 180.
25. Wallace, D.C. (1989) *Trends Genet.* **5**, 9.
26. Zeviani, M., Bresolin, N., Gellera, C. *et al.* (1990) *Am. J. Hum. Genet.* **47**, 904.
27. Hall, J.G. (1990) *Am. J. Hum. Genet.* **46**, 857.

Further reading

Davies, K. and Read, A.P. (1992) *Molecular Basis of Inherited Disease,* 2nd edn. Oxford University Press, Oxford.

Gelehrter, T.D. and Collins, F. S. (1990) *Principles of Medical Genetics.* Williams and Wilkins, Baltimore.

Weatherall, D.J. (1991) *The New Genetics and Clinical Practice.* 3rd edn. Oxford University Press, Oxford.

6
THE HUMAN GENOME: CLINICAL AND RESEARCH APPLICATIONS

6.1 Molecular dissection of common disease

Following the growing success of linkage analysis and positional cloning in mapping and isolating genes underlying single gene disorders, attention is increasingly being turned towards the molecular dissection of common disease. Many common human diseases do not show simple Mendelian inheritance but show evidence of genetic factors. For example, identical twins of type I diabetes patients have a sixfold higher risk of developing the disease than do other siblings.

The genetic contribution is multifactorial as a result of the additive effects of more than one gene locus, often perhaps involving only three to five major loci, and often with additional interactions with environmental factors. In studies of identical twins of individuals who have a common disease such as diabetes or schizophrenia, the other twin develops the disease no more than 20–50% of the time, suggesting the variable involvement of environmental factors (diet, smoking, viral infection, exposure to toxic agents, stress, etc.). In many cases, notably the common cancers, genes which have been shown to be implicated in rare inherited single gene disorders are increasingly being found to be important contributors to related common diseases.

Although the task of identifying susceptibility genes for common diseases is generally much more difficult than that of identifying disease genes for single gene disorders, the potential rewards are great. The identification of such genes may lead to a complete understanding of the mechanism of a common disease, and may elicit new or more effective treatments. The immediate benefits, however, lie in the realm of preventive medicine; early identification of individuals who have an increased risk of developing a disease may be followed by a planned program to reduce environmental factors which may precipitate the disease, regular clinical monitoring and early medical and surgical intervention.

6.1.1 Identifying disease susceptibility genes

In order to try to identify disease susceptibility genes, family-based linkage analyses (see Section 4.1.4) can, in principle, be used. The problem is that the lack of sensitivity of such methods in relation to multifactorial disease requires sampling of large numbers of families with a high incidence of the disease, the use of a large number of polymorphic markers covering the whole genome, and complex mathematical methods of multilocus analysis. However, progress is being made in several areas. A complete marker map of the human genome with an average spacing of 1 cM is expected in the next few years. In addition, considerable progress has been achieved in developing sophisticated mathematical methods to dissect complex human traits [1].

In the special case of identifying cancer susceptibility genes, a powerful additional approach has involved scanning for tumor-specific loss of constitutional heterozygosity for specific markers (see Section 6.1.2). In addition to the above methods, population-based association analysis can be employed to test the involvement of previously isolated genes which are regarded as candidate disease susceptibility genes. In this instance specific alleles of a candidate susceptibility gene are investigated in turn to see if any of them show a statistically significant association with the disease. An allele which shows significant association with the disease can then be regarded as a disease marker. The susceptibility to a disease that is conferred by a disease marker M, can be defined by a relative risk ratio (*Figure 6.1*).

Because the HLA loci are the most polymorphic human loci at the protein level, serologically defined alleles at different HLA loci have commonly been investigated in disease-marker association analysis, notably in the case of diseases with a known or suspected autoimmune component (*Table 6.1*). The recent characterization of multiple DNA-based polymorphisms at the HLA loci and a large variety of other loci has extended the scope of this type of analysis enormously. Once an apparently significant disease-marker association has been established, the extraordinary power of DNA technology can permit testing of the significance of these factors by creating transgenic animals (see Section 6.4) which contain the putative human disease susceptibility gene.

Another route to identifying human disease susceptibility genes requires identification and initial genetic analysis of an animal model of the human disease. Mapping of disease susceptibility genes to a specific chromosomal region in the animal can then suggest, by homology of synteny (Section 4.4.5), equivalent human chromosomal regions in which to concentrate the search for human disease susceptibility genes. The advantage of the animal model for mapping studies is that breeding crosses can be designed experimentally, unlike in human studies. Presently, the mouse genetic map is the most highly developed animal map and considerable information is available concerning homologous chromosomal regions in man and mouse.

	Affected individuals	Not affected
Number of individuals with allele m at marker locus M	a	b
Number of individuals who lack allele m at marker locus M	c	d
Relative risk of disease if allele m is present at marker locus M	$= \dfrac{ad}{bc}$	

Figure 6.1: *Relative risk calculation in disease-marker association analysis.*

Table 6.1: Examples of HLA-disease associations

| Disease | HLA marker | Percentage positive | | Relative risk |
		Patients	Controls	
Ankylosing spondylitis	B27 serotype	90	9	82
Rheumatoid arthritis	DR4 serotype	58	25	4.1
Insulin-dependent diabetes	DR3 serotype	46	22	3.1
	DR4 serotype	51	25	3.1
Narcolepsy	DR2 serotype	100	31	–
Celiac disease	DP RFLP	78	35	6.6

6.1.2 Cancer

Most cancers arise as a result of somatic mutation which is not inherited and therefore appears in sporadic cases. However, mutations in the germ-line can contribute to inherited cancers, including single gene disorders and common multifactorial cancers. It has been widely suspected that the development of common cancers is due to the combined effects of defective expression at a few disease loci. A powerful way of identifying important genes which contribute to common cancers involves the use of successive DNA markers in a two-pronged mapping strategy:

(a) loss of heterozygosity (Section 4.2.6) in common sporadic tumors;

(b) linkage analyses in rare familial counterparts of the common cancer.

The above strategy was adopted in the case of the common disease colorectal cancer, which has a rare familial counterpart, familial adenomatous polyposis (FAP) due to abnormal expression of a single gene. Following the identification of a cytogenetically visible deletion in 5q in a FAP patient, linkage analyses have mapped the *FAP* gene to 5q21 [2]. Subsequently, mapping of loss of tumor heterozygosity has revealed that the same chromosomal region is deleted in up to 60% of sporadic colon cancers suggesting that the *FAP* gene or a close neighbor is also involved in sporadic colon cancer. Recently, DNA probes which map to 5q21 were used to screen for somatic rearrangements in colon cancers, resulting in the identification of a new gene (*MCC* – mutated in colon cancer) which is implicated in colorectal cancer [3]. Subsequent analyses have, however, shown that FAP does not result from pathological mutations in the *MCC* gene. Instead, the *APC* gene, which is located in the chromosomal vicinity of *MCC*, has very recently been identified as the *FAP* gene by the demonstration of patient-specific mutations in the *APC* gene which are inconsistent with normal gene expression [4, 5].

The most commonly altered gene in human tumors is the *TP53* gene, normally a tumor suppressor gene but capable of being converted into a dominant oncogene in response to acquiring certain point mutations in its coding sequence [6]. Inherited

Table 6.2: *Genes which are implicated in common human cancers*

Gene	Chromosomal location	Nature of product	Involvement in common cancers
TP53	17p13.1	p53 tumor suppressor	75–80% of colon cancers show mutation of both alleles; 55% of primary breast cancers express mutant *p53*; also implicated in lung cancer and brain tumors
DCC	18q21	Tumor suppressor	Colorectal carcinoma: allele loss in 70% and absence or greatly reduced expression in 90% of such cancers
MCC	5q21	Tumor suppressor	Often deleted in colonic carcinoma cells?
MYC	8q24	Oncogene	Abnormal structure or function in lung cancer
RASA	5q13	*RAS* p21 oncogene	Abnormal structure or function in lung cancer
RAF1	3p25	Oncogene	Abnormal structure or function in lung cancer
JUN	1p32-p31	Oncogene	Abnormal structure or function in lung cancer
FER	5	Oncogene	Abnormal structure or function in lung cancer
ERBB2 (neu)	17q1-q2	Oncogene	Abnormal structure or function in lung cancer
Unidentified	17q21	?	Accounts for pathogenesis of a sub-set of familial breast cancer
Unidentified	17p13pter	?	Implicated in breast cancer
NM23	q21-q22	?	Suppresses metastasis

TP53 mutations are found in the rare dominantly inherited Li–Fraumeni familial breast cancer syndrome while somatic mutations appear to occur in a large fraction (perhaps even 50%) of all cancers. Although inherited *TP53* mutations are the initiating tumorigenic event in Li–Fraumeni Syndrome (where they occur 10–30 years before the onset of malignancy), alteration of the *TP53* gene is a rather late event in many common cancers. Tumorigenesis is therefore thought to result from a series of successive mutations in different cancer susceptibility genes in which the critical parameter is the cumulative effect of the mutations. The tumorigenesis of many common cancers appears to involve a variety of oncogenes and tumor suppressor genes (*Table 6.2, Figure 6.2*).

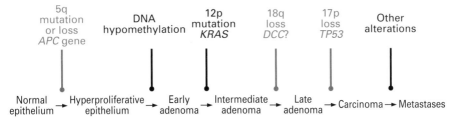

Figure 6.2: *Model for the interaction of oncogenes and tumor suppressor genes in colorectal tumorigenesis. Adapted from reference 7 with permission from Cell Press.*

6.1.3 Coronary heart disease

A strong correlation between high serum cholesterol levels and early coronary heart disease has long been recognized. Additionally, recent studies have demonstrated a direct association between the concentration of plasma fibrinogen and the subsequent incidence of ischemic heart disease and stroke; elevated fibrinogen levels may result in increased blood viscosity and an increased propensity for coagulation and thrombus formation. Identifying genes which contribute to common heart disease has not been facilitated by the large number of candidate disease genes which need to be investigated, including multiple genes involved in cholesterol and lipid metabolism and in blood clotting (*Table 6.3*). However, as in cancer studies, a lead has been given by the analysis of single gene disorders which are comparatively rare. Of these, the most significant is familial hypercholesterolemia, which occurs with a frequency of about 1 in 500, is dominantly inherited and is due to a defect in the LDL receptor gene on 19p. Because of abnormal expression of a single LDL receptor allele, heterozygotes have extremely high LDL concentrations and a high risk of having a heart attack. Defective expression of both alleles is found in rare homozyotes who often die of a heart attack before the age of 20.

In the last few years, considerable effort has been devoted to assessing whether specific alleles of candidate genes are markers of disease. From these studies it has been shown that certain haplotypes/alleles at apolipoprotein gene loci, for example, the *APOB* locus (2p), the *APOA1/APOC3/APOA4* cluster (11q) and the *APOC1/APOC2/APOE* cluster (19q), are correlated with elevated or reduced serum cholesterol levels. Additionally, marker-disease association analyses have also recently been successful in identifying individuals who are at increased risk of having a second heart attack because they have high concentrations of the clotting factor, fibrinogen. Such individuals often possess an RFLP in the regulatory region of the fibrinogen gene which is significantly associated with increased fibrinogen expression.

Recently, evidence from a linkage analysis study has suggested the existence of an LDL-suppressing gene [8]. Because of the possibility that this gene could protect against heart attacks, a considerable research effort is presently being directed to identifying it.

6.1.4 Diabetes

Of the two major forms of diabetes, type II is comparatively homogeneous genetically while type I, juvenile-onset, insulin-dependent diabetes is more heterogeneous. Type

Table 6.3: Human genes which are implicated in other common diseases

Disease	Implicated genes and chromosomal location	Comments
Coronary heart disease	Genes involved in lipoprotein metabolism e.g. LDL receptor (19p13) *APOB* (12p24-p23) and others	Mutations in LDL receptor genes cause familial hyper-cholesterolemia
	Clotting factors, e.g. fibrinogen? (4q28)	RFLP in regulatory region correlates with disease
		Gene closely linked to *HRAS* oncogene on 11p15.5, involved in rare long QT syndrome
Insulin-dependent diabetes mellitus	*HLA-DQA1, HLA-DQB1, HLA-DRB1,* all on 6p21.3	Certain alleles of these loci correlate with predisposition
	Insulin gene (11p15.5) Possible homologs of other unidentified murine susceptibility genes	Homologs expected on human chromosomes 1 (or 4), 2q and 17
Alzheimer's disease		
Early onset	Gene on chromosome 21 – *APP* (amyloid precursor protein)?	Involvement of genes on chromosomes 19 and 21 suggested by
Late onset	Genes on chromosome 21 and proximal 19q	linkage analyses. *APP* gene mutations correlate with disease in a small sub-set
Epilepsy	?	Two unidentified genes on chromosome 20 and 21 are closely linked to two rare forms of familial epilepsy
Asthma	Gene on chromosome 11q implicated in atopic immuno-globulin E responsiveness, the most common cause of asthma in children and young adults	

I diabetes appears to be an autoimmune disease in which perhaps 20–60% of the genetic component maps to the HLA complex, the human MHC located at 6p21.3. Certain alleles of at least three HLA genes, *HLA-DQA1*, *HLA-DQB1* and *HLA-DRB1*, whose function is to present peptide antigens to the T-cell receptor, are associated with predisposition to type I diabetes. In particular, comparative susceptibility or resistance to the disease is strongly correlated with the nature of amino acid number 57 in the HLA-DQ$_\beta$ chain, which is a component of a helix that defines part of the putative antigen presenting site; aspartate is negatively correlated with the disease whereas substitution by alanine, serine or valine is often found in patients [9].

In the spontaneous non-obese diabetic (NOD) mouse mutant, an animal model of type I diabetes, a similar aspartate to serine substitution is found at position 57 in the analogous mouse *I-A* gene. However, the single Asp→Ser substitution in the I-A^{NOD} allele is not sufficient for the development of the disease as the expression of experimentally introduced normal *I-A* genes containing either Asp-57 or Ser-57 can prevent the development of insulitis in NOD mice [10]. In addition, examples are known of human type I diabetics with HLA-DQ$_\beta$ molecules containing Asp-57. Accordingly, some HLA-DQ$_\beta$ molecules appear to confer various degrees of resistance to disease while others are neutral [11].

The insulin gene region on 11p is also known to be associated with diabetes susceptibility. Other non-HLA human diabetes susceptibility genes remain to be identified. Analysis of the NOD mouse has suggested that in addition to the *idd-1* locus, which maps to the murine MHC, there are at least three other murine diabetes-susceptibility loci: *idd-3*, *idd-4* and *idd-5*, which would be expected to have human homologs on chromosome 3, chromosome 1 or 4, and chromosome 2q, respectively [12, 13].

6.1.5 Dementia and mental illness

Alzheimer's disease is a group of disorders which collectively constitute the most common age-related cerebral diseases. Heritability is higher for early onset forms and there is a rare autosomal dominant form of familial Alzheimer's disease. Linkage studies have emphasized a high degree of genetic heterogeneity; certain early onset families appear to show linkage with chromosome 21 markers whereas late onset (> 65 years) families have also shown evidence of linkage to markers on 19q [14, 15]. The precursor protein of the β-amyloid deposited in the brains of Alzheimer patients is known to be encoded by a gene, *APP*, which is located at chromosome 21q21.2 in a region consistent with the early onset linkage finding. Recently, the same mutation has been found in the *APP* gene of affected members of two families with early Alzheimer's disease [16] and has subsequently been reported in French and Japanese families. Although this patient-specific mutation induces a conservative amino acid change (valine to isoleucine), the mutation occurs in a critical region of the *APP* gene, suggesting that it could be directly involved in the pathogenesis of a small subset of Alzheimer patients. Follow-up research is investigating other possible patient-specific mutations in the *APP* gene in Alzheimer families where there is evidence of linkage with chromosome 21 markers.

Recently, genetic linkage has been reported between chromosome 11 markers and manic depression in a large Amish family and between schizophrenia and chromosome 5 markers in one study looking at Icelandic and British families. However, subsequent failure to confirm these results has cast strong doubts on the validity of the original findings. Part of the problem with these diseases is that accurate diagnosis remains extremely difficult and misdiagnosis of only a small number of individuals is sufficient to alter the linkage result profoundly.

6.2 Studying gene expression and function at disease loci

Because of the success of 'reverse genetics', it has been possible to investigate the structure, expression and function of genes at a variety of previously uncharacterized disease loci. Structural characterization of normal and disease alleles permits a study of the molecular pathology of the disorder and permits the development of accurate DNA-based diagnosis. Computer analysis of the DNA sequence and predicted polypeptide sequence (searches for structural motifs, homologies with other previously reported sequences) can provide clues as to the function of the product. Additionally, a variety of expression studies and functional assays involving normal, disease, and experimentally manipulated alleles can provide insights concerning how the normal allele functions, the range of tissues in which it is expressed and how its defective expression can cause disease. Expression studies have also led to characterization of the normal gene product, permitting novel diagnostic assays.

In the following sections examples of these approaches are described with specific reference to the cystic fibrosis and Duchenne/Becker muscular dystrophy loci. The knowledge obtained from the above kinds of study will hopefully lead to the development of new or improved treatments for disease (see Section 6.5).

6.2.1 The cystic fibrosis transmembrane regulator gene

Cystic fibrosis is a recessive disorder characterized primarily by viscous secretions of the lungs and pancreas leading to a variety of symptoms including chronic lung disease. Approximately 1 in 20 Caucasians is a carrier and about 1 in 2000 live births is affected. The primary defect in cystic fibrosis patients is now known to be abnormal regulation by cAMP of chloride ion secretion by epithelial cells; inadequate secretion of Cl⁻ is believed to cause the insufficient hydration of mucus in the airways and pancreatic ducts.

The *CFTR* gene contains 27 exons and is known to span about 250 kb. At the time of writing approximately 10% of the gene had been sequenced including all exons and exon–intron boundaries [17]. The DNA sequence analysis predicts an encoded polypeptide of 1480 amino acids which is organized in five domains: two ATP-binding domains, two membrane-spanning domains (each with six transmembrane regions)

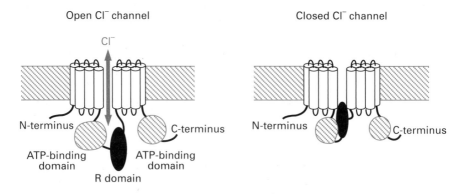

Figure 6.3: *Model for CFTR function and regulation.*

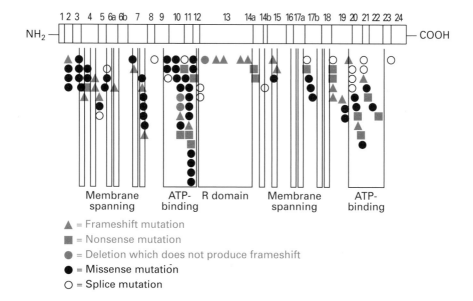

Figure 6.4: *Cystic fibrosis mutations. Numbered boxes represent segments of the CFTR product encoded by exons 1–24.*

and the R domain, which is rich in charged amino acids and potential sites for phosphorylation (*Figure 6.3*).

The structure of the *CFTR* gene suggests strong parallels with a family of proteins which actively transport specific molecules in an ATP-dependent fashion. However, recent studies suggest that the *CFTR* gene encodes a simple Cl⁻ channel; when the *CFTR* gene is experimentally introduced into and expressed in non-epithelial cells, the property of cAMP-regulated Cl⁻ ion conductance is conferred on the recipient cells [18,19]. Most probably the transmembrane domains form a channel for Cl⁻ ion transport across epithelial cell membranes, and the R domain opens and closes the channel in response to cAMP-activated protein kinases (*Figure 6.3*). As yet, the function of the ATP-binding domains is unknown. In Caucasian populations approximately 45–80% of cystic fibrosis disease alleles are accounted for by a single mutation, ΔF508, in which a 3-bp deletion in exon 10 results in removal of a codon specifying phenylalanine. However, by mid July 1991, the international cystic fibrosis genetic analysis consortium had reported that in addition to the major ΔF508 mutation, over 100 different minor pathological mutations contribute to cystic fibrosis, each occurring at very low frequencies. Nearly 50% of all pathological mutations, including the ΔF508 mutation, appear to be clustered in the ATP-binding domains (*Figure 6.4*).

6.2.2 The dystrophin gene

Duchenne and Becker muscular dystrophy result from pathological mutations in the dystrophin gene which is about 2.3 Mb long and has more than 100 exons, collectively accounting for 14 kb or about 0.6% of the gene. The complete nucleotide sequence of the exons has been obtained from DNA sequencing of overlapping dystrophin cDNA clones [20]. The inferred protein product appears to be rod-like in shape and has 3685 amino acids. It has been considered to be organized in four domains:

(a) a 240-amino acid N-terminal domain which shows homology to the actin-binding domain of α-actinin;

(b) a large central domain which has 24 diverged repeats of a roughly 109-amino acid unit, showing weak similarity with similar-sized repeats found in certain cytoskeletal proteins such as spectrin;

(c) a cysteine-rich domain which shows some similarity to the C- terminal domain of α-actinin;

(d) a C-terminal domain which is not related to other molecules other than a protein encoded by a dystrophin-like gene, *DMDL*, on 6q24 [21] (see *Figure 6.5*).

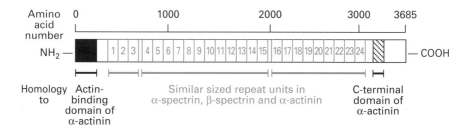

Figure 6.5: *Domain structure of dystrophin. Numbered boxes represent internal repeat units.*

From RNA and immunochemical assays the dystrophin gene is known to be significantly expressed in four human tissues: skeletal muscle, cardiac muscle, smooth muscle and brain. Alternative promoters and first exons characterize the expression of the dystrophin gene in muscle and brain (see *Figure 2.14*). Additionally, the 3' end of dystrophin mRNA can undergo alternative splicing to generate isoforms at the C-terminus. Presently, the exact biological function of dystrophin is unknown but it is known to be a cytoskeletal membrane protein which interacts with several other membrane proteins and is localized to the cytoplasmic face of the cell membrane. An initial stage in the molecular pathogenesis of DMD has been speculated to be the loss of a dystrophin-associated glycoprotein which is observed to accompany loss of dystrophin expression [22].

6.3 Diagnostic applications

6.3.1 Prenatal diagnosis and pre-morbidity testing

The availability of gene probes and/or closely linked DNA markers for a large variety of single gene disorders permits predictive testing for the inheritance of a disease gene or its normal allele at early stages in fetal development. Such early diagnosis permits

Table 6.4: *Selected examples of inherited disorders for which DNA-based prenatal tests have been used or are feasible.*

Adult polycystic kidney disease	21-hydroxylase deficiency
Agammaglobulinemia	Hypercholesterolemia
α1-antitrypsin deficiency	Hyperlipidemia
Antithrombin III deficiency	Lesch–Nyhan Syndrome
Choroideremia	Marfan Syndrome
Chronic granulomatous disease	Neurofibromatosis type I
Cystic fibrosis	Neurofibromatosis type II
Duchenne/Becker muscular dystrophy	Ornithine transcarbamylase deficiency
Fragile X-linked mental retardation	Osteogenesis imperfecta
Gaucher's disease	Phenylketonuria
Growth hormone deficiency	Retinoblastoma
Hemophilia A	Sickle cell anemia
Hemophilia B	α-thalassemia
Huntington's disease	β-thalassemia

the possibility of selective termination of pregnancies in which the fetus is diagnosed to carry disease genes. In the case of diseases where the age of onset is relatively late, it can permit targeting of early surgical or medical intervention to minimize subsequent morbidity. Following the initial report of DNA-based diagnosis more than a decade ago, there was initially slow progress. However, an ever increasing number of identified disease genes and linked markers together with recent technological developments, have led to subsequent rapid developments (*Table 6.4*) and widespread applications [23, 24].

Two basic approaches are possible in DNA-based diagnosis. If the appropriate gene probes are not available, then indirect analyses are required, involving linkage analyses in which the test sample is normally referenced against DNA samples from family members, including the parents and a previously affected individual. DNA-based linkage analyses employ extragenic flanking DNA markers, and, when available, intragenic DNA polymorphisms. For each informative marker the theoretical accuracy of the test is dictated by the recombination fraction between the marker and the disease locus. For intragenic DNA polymorphisms, the recombination fraction between the marker locus and the disease locus is usually about zero but may be a few per cent in the case of giant genes such as the dystrophin gene.

More distantly located extragenic DNA polymorphisms inevitably admit higher error rates due to recombination between the marker and disease loci. However, the availability of panels of DNA markers which are located centromeric to, and also telomeric to the disease locus, so-called flanking markers, can permit extremely accurate diagnosis. The hypothetical example in *Figure 6.6* illustrates a problem with predictive testing for Huntington's disease at the time of writing. It is probable, but not proven, that the D4S125 and D4S96 markers flank the disease locus. If so, it can be inferred that the Huntington's disease allele is being transmitted in this pedigree on a chromosome in which the flanking marker loci D4S125 and D4S96 show alleles 3 and 2, respectively. The recombination fraction between either of these probes and Huntington's disease can conservatively be estimated to be less than 2%. Therefore, the genotyping results for the daughter of the affected man would predict with great accuracy (somewhat greater than 99.96%) that she has not inherited the disease allele (the small error rate is due to the extremely unlikely possiblity of a double cross-over).

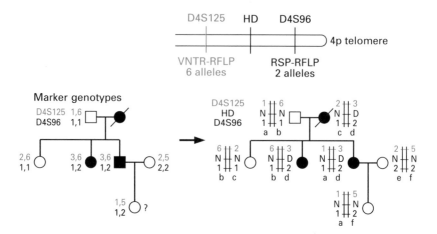

Figure 6.6: *DNA-based predictive testing in a Huntington's disease family.*

However, because of the uncertainty concerning whether these markers really do flank the HD locus, we are left with the much higher error rate in diagnosis due to the possibility of recombination between the disease locus and an individual marker.

If gene probes are available to permit detection of the disease mutation, then direct analyses may permit identification of a disease gene in an individual DNA sample, without the need to reference the sample against those of other family members. If the disease is known to be genetically homogeneous (e.g. sickle cell disease), or if there is prior information from other affected family members concerning the nature of the specific mutation, various analyses can be used to identify the disease mutation directly. However, individual disease loci are often associated with a high degree of mutational heterogeneity. In such cases, if the nature of the specific mutation in the relevant previously affected family members is unknown, it is often more practical to use indirect linkage analyses rather than attempt to screen for a variety of different possible pathological mutations.

DNA-based prenatal diagnosis has generally been conducted at the chorion villus level at the 9th–12th week of gestation, which has the advantage of affording earlier diagnosis than traditional amniocentesis sampling (often conducted at 15–17 weeks of gestation). However, the extraordinary sensitivity of the PCR means that PCR-based analyses can be conducted to detect a disease allele in a single human cell [25]. Consequently, it is now feasible to remove a single cell from a pre-implantation embryo in order to diagnose genetic disease. The sensitivity of PCR-based detection methods, however, affords the possibility of false diagnosis due to unwanted amplification of contaminant cells or tissue. In the case of amniocentesis sampling where the fetus is female, the accuracy of interpretation of PCR-based analyses can be threatened by maternal cell contamination.

6.3.2 Carrier testing

Family-based carrier testing can be conducted by linkage analyses in families in which there is a surviving affected individual. Except in the case of rare *de-novo* mutations,

Exons	bp
48	506
44	426
51	388
43	357
45	307
50	271
53	212
47	181
42	155
60	139
52	113

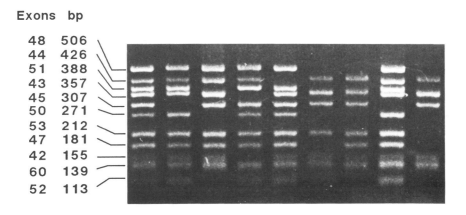

Figure 6.7: *Detection of dystrophin gene deletions by multiplex PCR amplification of individual exons. Reproduced from reference 26 with permission from the British Medical Association.*

parents of a child with an autosomal recessive disorder are carriers of the disorder, with one mutant gene and one normal allele at the disease locus. Assuming an early age of onset for the recessive condition, unaffected siblings of an affected child have a 2/3 risk of being a carrier.

Indirect linkage analyses always carry an error rate due to recombination between the marker and the disease site. In some cases, this error rate may be appreciable even though gene probes are available. For example, DMD has a high degree of mutational heterogeneity (due to intragenic deletions in about 65% of cases) and the dystrophin gene shows a very high degree of intragenic recombination. Accordingly, it is highly desirable to be able to identify the pathological mutation directly in an affected child in order to permit accurate prediction of carrier status of affected female relatives and in subsequent prenatal diagnostic testing. Recently, multiplex PCR methods have been developed which permit direct screening for 98% of Duchenne/Becker muscular dystrophy deletions. Such tests involve multiple PCR assays using primers which flank specific exons and simultaneous assay by agarose gel electrophoresis of multiple differently sized exon-specific PCR products (*Figure 6.7*).

Population-based carrier screening has also been contemplated for certain common single gene disorders. In such cases, direct testing for disease mutations is needed and a knowledge of the full extent of mutational heterogeneity is required to permit testing for the presence of different disease alleles. Although certain disorders are mutationally heterogeneous, the distribution of pathological mutations in specific populations may be rather limited, facilitating carrier screening.

Cystic fibrosis, predominantly a disease of Caucasians, was expected to be mutationally quite homogeneous on the basis of linkage disequilibrium (see Section 5.1.4). In support of this view a single mutation, DF508, accounts for the majority of present-day cystic fibrosis alleles. In addition, however, there are numerous minor cystic fibrosis alleles (see *Figure 6.4*).

The ΔF508 mutation can easily be screened by directly size-fractionating PCR products using primers which flank the mutation site (*Figure 6.8a*). Additionally, PCR-based ARMS assays (see Section 3.3.3) can be used to screen for other mutations such as G542X, a G→T substitution in exon 10 which replaces a GGA codon (glycine) by a stop codon, TGA. In the example illustrated in *Figure 6.8b*, two PCR reaction assays are conducted with one common conserved upstream primer and a second allele-specific primer which binds to a region that includes the mutation site and the DNA region immediately downstream. The 3' nucleotide of the allele-specific primer is either A (G542X-specific) or C (specific for the normal allele) and successful amplification is possible only when the 3' end nucleotide is matched with its complementary nucleotide.

Despite the convenience of PCR assays for cystic fibrosis mutations, the sheer number of tests that would have to be conducted for each individual presently make DNA-based population screening for cystic fibrosis carriers an unattractive proposition.

6.3.3 Predictive testing based on analysis of gene product

Positional cloning has permitted the isolation of human disease genes such as the dystrophin gene and the *CFTR* gene without prior knowledge of the gene product. Subsequent characterization of such genes has permitted definition, for the first time, of their gene products. Because of the difficulties which are sometimes associated with identifying mutations at the gene level (see previous sections), the alternative novel possibility of analyzing the relevant gene products can be investigated. In the case of large genes it may be much simpler to study the mRNA product by PCR-based analyses. Even if the mRNA for such a gene is normally expressed in a tissue which is difficult to access, the phenomenon of illegitimate transcription means that a small amount of the relevant mRNA will be expressed in blood cells and can be amplified by PCR to generate enough material for study. Recently, for example, it has been possible to amplify the entire coding sequence of dystrophin mRNA from blood cells, which permits detection of intragenic deletions and duplications in DMD patients and positive diagnosis of carriers with such mutations [27].

6.3.4 Gene mapping as an aid to clinical diagnosis

Certain groups of human genetic disorders show a range of similar or overlapping phenotypes. In some cases different disease genes are responsible for producing very similar clinical phenotypes and this can be an important source of error in DNA-based diagnosis. Conversely, if different clinical phenotypes can be shown to be due to mutation at the same disease locus the same set of DNA probes can be employed in predictive testing for apparently different diseases. Linkage studies using DNA markers have therefore proved to be an important approach to identifying possible

Figure 6.8: *PCR-screening for cystic fibrosis mutations ΔF508 and G542X.*
(a) Conventional PCR assay for ΔF508 on 12 samples reveals two ΔF508 homozygotes (lanes 2 and 8) and five heterozygotes (lanes 3, 4, 5, 7 and 10).
(b) PCR–ARMS assay for G542X reveals two heterozygotes (lanes 2 and 4). To ensure that the DNA in individual samples is amenable to PCR amplification, a control PCR reaction is conducted using two conserved CFTR-specific amplimers.

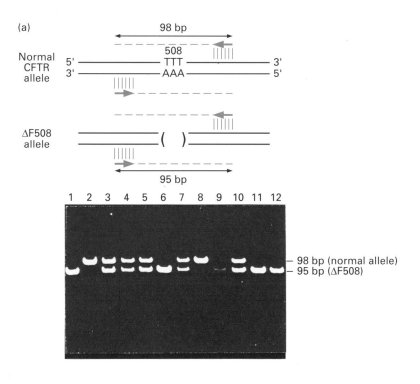

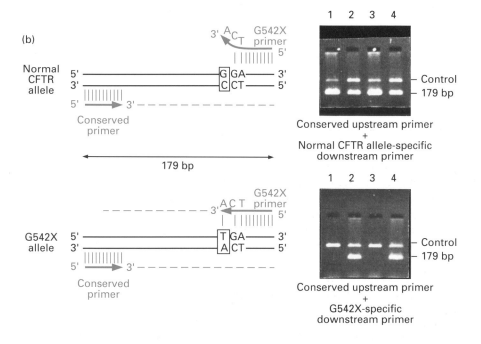

disease locus heterogeneity and in dissecting different related genetic disorders. In some cases such approaches have facilitated or confirmed clinical diagnosis.

Disease locus heterogeneity. In certain genetic diseases locus heterogeneity is apparent from different modes of inheritance. For example, retinitis pigmentosa can be transmitted by autosomal dominant, autosomal recessive or X-linked recessive modes of inheritance. The first evidence for disease locus heterogeneity using linkage to polymorphic markers was obtained in the case of elliptocytosis. Two genetically distinct varieties of this dominantly inherited disorder can be distinguished by the observation that one form is genetically linked to the Rhesus locus on chromosome 1, whereas the other is not. The advent of abundant polymorphic DNA markers has resulted in the identification of a number of parallel cases (*Table 6.5*).

Table 6.5: *Disease locus heterogeneity*

Disease	Locus	Chromosomal location	Mode of inheritance[a]
Charcot–Marie–Tooth	CMT1A	17p13.1-p11.1	AD
neuropathy	CMT1B	1q	AD
	CMTX	Xq11-q13	XLR
Neurofibromatosis	NF1	17q11.2	AD
	NF2	22q11-q13.1	AD
Polycystic kidney disease	PKD1	16p13.3	AD
	PKD2	?	AD
Retinitis pigmentosa	RP1	1	AD
	RP2	Xp11.4-p11.21	XLR
	RP3	Xp21.1	XLR
	RP5	3q	AD
	RP6	Xp21.3-p21.2	XLR
Tuberose sclerosis	TSC1	9q	AD
	TSC2	11q14-q23	AD
	TSC3	12q	AD
Usher's Syndrome	USH1	?	AR
	USH2	1q	AR
Waardenburg Syndrome	WS1	2q35	AD
	WS2	?	AD

[a]Autosomal dominant, AD; autosomal recessive, AR; X-linked recessive, XLR.

Disease locus homogeneity. Gene mapping, isolation and characterization studies have demonstrated that certain disease phenotypes which were previously considered to be possibly genetically distinct, are in fact allelic variants. In some cases disorders which are clinically heterogeneous such as Friedrich's ataxia and spinal muscular atrophy have shown no evidence of locus heterogeneity. Extended linkage analyses in large numbers of families can be used to confirm locus homogeneity, as in Huntington's disease.

6.4 Creating animal models of human disease

Animal models of human genetic disease are important to research because they allow detailed examination of the pathophysiology and may permit the design of novel treatments of such disorders. A number of spontaneous animal mutants are known. In some cases the animal mutant phenotype closely parallels the corresponding clinical phenotype, but in others there is considerable divergence because of species differences in biochemical and developmental pathways. For example, the mouse *mdx* mutant has a very mild phenotype compared to the homologous DMD in humans. Recombinant DNA technology now offers the possibility of creating specific animal models of human genetic disease, provided that the underlying human or animal gene has been identified. Essentially two approaches can be used: (a) transfer of a human disease gene into the germ-line of an animal; (b) specific inactivation of the homologous animal gene in embryonic cells (*Table 6.6*; see below).

Table 6.6: *Examples of transgenic animal models of human disease*

Disorder	Method of creating transgenic animal model
Ankylosing spondylitis	Microinjection of human genes for HLA-B27 and associated light chain, β_2-microglobulin into rat embryos
DiGeorge's Syndrome	Specific inactivation by gene targeting of the *hox-1.5* gene in mouse embryonic stem cells
Gerstmann–Straussler–Scheinker Syndrome and Creutzeld–Jakob disease	Microinjection of artificially mutated mouse *PRNP* gene into fertilized mouse oocytes
Retinoblastoma	Microinjection of a monkey viral oncogene, SV40, T antigen into fertilized mouse oocytes. Expression of the viral oncogene in retinal cells results in inactivation of the normal retinoblastoma tumor suppressor gene and the consequent development of ocular tumors
Sickle cell disorder	Microinjection into fertilized mouse eggs of a construct containing human α-globin genes, the human β^s-globin gene and the β-globin LCR region
Stomach cancer	Microinjection into fertilized mouse eggs of human adenovirus oncogenes

Most experiments have involved producing transgenic mice. Direct transfer of a human or mouse disease gene into the mouse germ-line is possible by microinjection of the cloned disease gene into a fertilized mouse oocyte (*Figure 6.9*). A certain proportion of the introduced DNA molecules become stably integrated into the mouse genome, and expression of the disease allele can result in an altered phenotype which may serve as an animal model of human disease. In the case of a dominantly inherited disease, the pathogenic potential of a single introduced disease allele would not be expected to be overcome by the expression of normal alleles at the corresponding mouse locus.

In some cases the animal homolog of the human disease gene can be introduced into the mouse germ-line to verify whether a candidate mutant human gene is responsible for pathogenesis. For example, specific mutations in the human prion protein gene (*PRNP*) have been implicated in the pathogenesis of Gerstmann–Straussler–Scheinker Syndrome and Creutzfeld–Jakob disease, disorders of brain function whose pathogenesis is similar to that of scrapie, a degenerative brain disease in sheep. In order

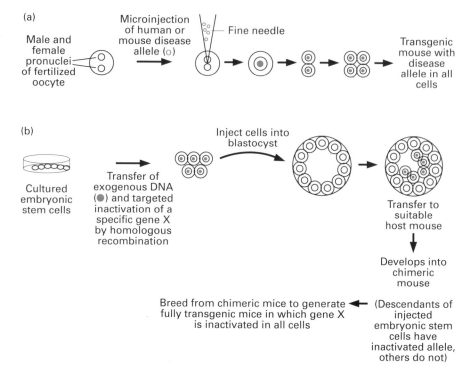

Figure 6.9: *Creating transgenic mouse models of human disease. (a) Microinjection of disease allele into fertilized mouse cocytes. (b) Gene inactivation by homologous recombination in embryonic stem cells.*

to test whether mutations in the *PRNP* gene cause the pathology, a transgenic mouse was created by injection into the mouse embryo of a mutated mouse *PRNP* gene which was directly homologous to a human *PRNP* mutant found in Gerstmann–Straussler–Scheinker Syndrome. The resulting transgenic mouse showed spontaneous neurodegeneration with reproduction of many of the clincal and pathological features of Gerstmann–Straussler–Scheinker Syndrome. [28].

Recently, transgenic animals have also been created to test highly significant marker-disease associations. For example, the human *HLA-B27* gene which is significantly associated with the arthropathy, ankylosing spondylitis, has been incorporated into the genome of a rat embryo and the resultant transgenic rat has spontaneously developed inflammatory disease, thereby providing an animal model of HLA-B27-associated disease [29].

A mouse transgenome can also be created by transfection of foreign DNA into cultured mouse embryonic stem cells. The embryonic stem cell approach affords the potential for gene targeting by homologous recombination *in vitro (Figure 6.10)*. Using this procedure it is possible to inactivate a specific predetermined gene in the mouse genome by recombination with an introduced gene. Subsequently, embryonic stem cells can be implanted into the blastocyst of a pseudopregnant mouse. Because some of the transfected embryonic stem cells can give rise to germ-line cells, the resulting progeny can be screened to identify mice that have inherited the inactivated gene. Mice

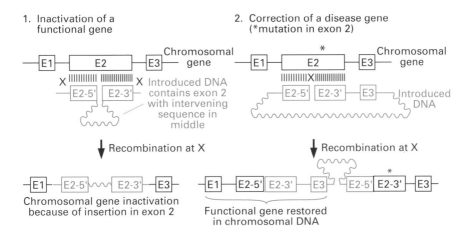

Figure 6.10: *Gene targeting by homologous recombination.*

which are carriers of the desired null mutation can then be interbred to generate offspring which lack a specific gene product. The resulting phenotype may therefore serve as a model of a human disease which is due to genetic deficiency at the homologous locus.

In some cases, targeted inactivation of a mouse gene may serendipitously produce a mouse model of human disease. For example, targeted inactivation of the mouse homeobox gene *hox-1.5* has generated a mouse phenotype which is remarkably similar to DiGeorge's Syndrome and may provide some kind of animal model for this disease [30].

6.5 Treatment of genetic disease

The 1980s witnessed a remarkable upsurge in the application of DNA technology to the diagnosis of disease and at the same time a rapid increase in our understanding of the molecular basis of disease. Increasingly in the 1990s attention will be focused on the application of DNA technology to the treatment of disease.

6.5.1 Cloned human genes as a source of medically important products

Treatment of a number of human diseases has used biochemically purified animal or human products to supplement a deficiency of a human protein. For example, diabetes sufferers are often treated with insulin prepared from cows or pigs. However, because of species differences in the amino acid sequence of the product, animal products are potentially immunogenic and may produce unwanted side-effects in highly immunoreactive individuals. The administration of biochemically purified human products may also be hazardous. Recently, many hemophiliacs have developed AIDS as a result of treatment with Factor VIII which was purified from the serum of unscreened human donors.

Isolation of the human gene and expression in suitable expression systems should permit large quantities of the human product without the attendant hazards of the type

Table 6.7: *Human proteins of therapeutic value which are commercially produced by genetic engineering*

Protein	Potential/realized therapeutic value
Blood clotting Factor VIII	Treatment of hemophilia A
Interferon	Anti-cancer agent; treatment of hepatitis B
Interleukin-2	Suppression of autoimmune disease, enhancer of graft survival in heart transplantation
Erythropoietin	Treatment of anemia
Growth hormone	Treatment of growth hormone deficiency
Tissue plasminogen activator	Treatment of coronary thrombosis

described above. Recently, a number of cloned human gene products of medical interest have been produced on an industrial scale (see *Table 6.7*). Often these products have been produced by cloning and expressing the human gene in micro-organisms. In such cases the expressed human protein may undergo post-translational modification (e.g. glycosylation) which is different to that undergone in human cells and is potentially immunogenic. To cirumvent these difficulties, increasing attention has been paid to constructing transgenic livestock whose post-translational processing systems are more similar to analogous human systems. For example, a cloned human gene can be fused to a sheep gene specifying a milk protein and then inserted into the genome of the sheep germ-line. The resultant transgenic sheep can secrete large quantities of the fusion protein in its milk. The design of the fusion gene normally allows the secreted fusion protein to be cleaved using a specific protease in order to generate the human protein residue.

6.5.2 Gene therapy

Gene therapy, the artificial introduction of genes into disease tissue in order to overcome disease, has recently been initiated in the case of certain human diseases (see below). Where possible, it is desirable that the introduced genes become integrated into the chromosomal DNA of recipient cells and therefore have the potential of being expressed in descendants of the original cells, following cell division. A variety of different strategies have been considered, of which the simplest is to transfer the desired genes into appropriate cells in culture then insert the transfected cells into the patient.

Presently, attention is being focused on somatic gene therapy, in which specific tissue types are targets for gene therapy [31]. Germ-line gene therapy is also potentially feasible. In the case of familial disease, it is theoretically possible to identify an embryo as carrying disease genes and to apply gene therapy in order to prevent the expression of disease. However, although successfully carried out in animal studies, germ-line gene therapy has not been considered in human studies. By manipulating the DNA of the embryo, the germ-line will be modified so that the consequences of the therapy are not restricted to the treated individual but may also be applicable to descendants of the treated patient. Additionally, current procedures for transfecting exogenous DNA into embryos are extremely inefficient and liable to serious error. In particular, there is little control over the location of the chromosomal sites into which the introduced DNA integrates. The resulting virtually random integration of introduced DNA into chromosomal DNA can therefore be pathological; insertional inactivation of an important gene may ensue while unwanted activation of an oncogene could lead to tumorigenesis. In any case, even in the most serious inherited disorder, generally at least 50% of the embryos will be normal and can be identified as such by sensitive PCR-based methods (see Section 6.3.1).

Gene augmentation versus gene correction. Obvious targets of somatic gene therapy include single gene disorders where the underlying gene has been well characterized. Two types of strategy are required. In the case of recessive disorders where the disease is due to absence of a normal gene product, theoretically it is sufficient to introduce a functioning normal allele of the relevant gene in order to overcome the genetic deficiency (gene augmentation). However, in the case of dominant disorders the pathogenic potential of the mutant allele is normally expressed in the presence of a normal allele. Consequently, gene therapy for dominant disorders requires gene correction, that is, replacement of the mutant sequence by the equivalent sequence from a normal allele, or selective inactivation of the mutant gene. Such procedures are more difficult as the only approach to achieving them at present is gene targeting by homologous recombination (*Figure 6.10*). Recently, it has been possible to achieve correction of a human β^S-globin allele to the normal β^A-globin allele in cultured cells [32]. However, at present, gene correction is a very inefficient procedure and, consequently, initial attempts at gene therapy have concentrated on recessive disorders.

Delivery to target tissue. In many cases the normal allele of a disease gene may be expressed in a tissue that is difficult to access, such as brain or liver. However, if the deficiency is in a gene which encodes a secreted protein, then it may be possible to design a gene therapy whereby the normal gene is integrated into the chromosomal DNA of a more accessible tissue than the one in which the gene is normally expressed. The most accessible tissues are blood cells and, to a lesser extent, skin fibroblasts. Hematological disorders are therefore prime candidates for gene therapy. In principle it is possible to remove blood cells from a patient, grow them in culture, transfect the desired gene into the cultured cells, select for suitably transfected cells, then transfuse these cells into the patient. However, blood cells are continually dying and being replaced. The long term effectiveness of gene therapy for hematological disorders therefore depends on the proportion of blood stem cells in which gene therapy has been successful. Unfortunately, as yet, it has not been possible to isolate human blood stem cells.

Transfer of genes into cultured human cells is most effectively carried out using vectors based on retroviruses. These are RNA viruses which can infect human cells with a high efficiency and which, following infection, undergo conversion to a double-stranded DNA form that integrates into chromosomal DNA. Normally only one to two copies of the retrovirus DNA integrate into the genome of recipient cells, so that the potential for pathological effects following integration (inactivation of a resident gene, activation of an oncogene, etc.) are restricted. Using standard DNA technology, a human gene can be inserted into the double-stranded DNA form of a retrovirus and the recombinant retrovirus can be packaged into a protein coat, suitable for high level infection, by the provision of packaging proteins supplied by helper retroviruses.

Expression. In some cases the level of gene expression is crucially important. For example, β-thalassemia was initially considered a prime candidate for gene therapy. The β-globin gene is very small (1.6 kb) and all the important signals required for high level expression of this gene have been identified and assembled in a mini-gene construct. However, although the human β-globin gene can be expressed at a high level, the regulation of expression is complex because it it is co-ordinated with α-globin gene expression. Attempted gene therapy of β-thalassemia must therefore ensure expression of β-globin at the correct level. If it is too high there will be an imbalance of globin chains with a relative deficiency of α-globin, which can provoke α-thalassemia.

Current gene therapy. The first serious attempt at gene therapy for a single gene disorder has focused on adenosine deaminase (ADA) deficiency, a rare recessively inherited disorder of purine metabolism. Patients with this defect account for 25% of all cases of severe combined immunodeficiency in which the immune system is compromised prinicipally because of ADA deficiency in T lymphocytes. These cells are relatively accessible. Additionally there is a wide range in the levels of ADA expression in normal individuals, suggesting that the precise regulation of expression of this gene is not essential. In animal experiments, equivalent treatment of immune-deficient mice by administration of ADA alone did not lead to long-term immune reconstitution. However, mice whose cells had been transfected with the human *ADA* gene showed normal immune function for up to 3 months. Subsequently, gene therapy

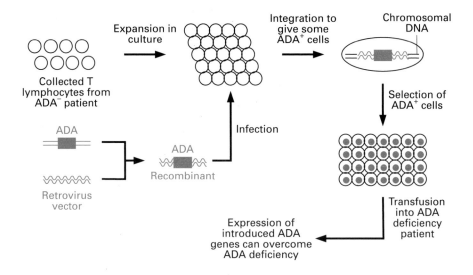

Figure 6.11: *Gene augmentation therapy for adenosine deaminase (ADA) deficiency.*

for ADA deficiency was approved and initiated by researchers at the U.S. National Institutes of Health (N.I.H.) on 14 September, 1990, according to the experimental plan illustrated in *Figure 6.11*. Clearly the envisaged treatment is not a permanent cure and requires repeated injections with *ADA*-gene-transfected short-lived T cells. Although the degree of success of the experiment will not be known fully for some time, early reports appear encouraging.

6.5.3 Cancer gene therapy

The first approved human gene therapy for a cancer was initiated by researchers at the N.I.H. and involves an attempt to treat the comparatively common cancer, malignant melanoma, which is incurable in its final stages. The design of this experiment involves the use of autologous tumor-infiltrating lymphocytes (TIL) as vehicles for gene transfer. These lymphoid cells infiltrate solid tumors and can be grown *in vitro* by culturing single cell suspensions from tumors in the presence of interleukin 2. An initial phase of this experiment, and the first recorded case of gene transfer into humans, was the introduction of a marker bacterial neomycin gene, by retrovirus-mediated transfer, into TIL in order to monitor the fate of the TIL [33]. There were no side-effects attributable to the procedure, and approval has recently been given to a gene therapy protocol involving retrovirus-mediated transfer of a gene encoding tumor necrosis factor into TIL from a melanoma patient (*Figure 6.12*). In this procedure suitably transfected TIL are injected into patients where they are expected to home into even deep-seated tumors and to express tumor necrosis factor, hopefully causing the tumors to regress.

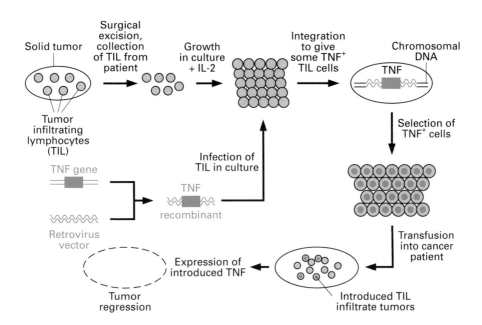

Figure 6.12: *Cancer gene therapy by adoptive immunotherapy. TNF, tumor necrosis factor.*

6.5.4 Future prospects

Over the last decade editorials on the prospects of gene therapy have oscillated between optimism and pessimism in response to a series of technical breakthroughs and setbacks.

As the *in-vitro* expression of more and more disease genes and their normal alleles are being studied, an increasing number of gene therapy experiments are now being contemplated, including examples of disorders such as DMD and cystic fibrosis in which the idea of gene therapy was formerly considered quite unrealistic. In the former case there is the problem of a giant gene being expressed in a relatively inaccessible tissue, muscle. However, the coding sequence of the dystrophin gene is only 0.5% of the gene size, about 14 kb, and a significant amount of this may be deleted without severe clinical consequences (see Section 5.5.1). Consequently, it is possible to engineer a truncated minigene construct containing the remainder of the dystrophin coding sequence which is of a size that can be accepted by retrovirus vectors. Delivery, however, remains a problem, although the possibility of delivery of a minigene construct into muscle fibers via intermediate muscle satellite cells has been considered.

In the case of cystic fibrosis, aerosol-mediated delivery of the *CFTR* gene to airway epithelial cells of the lungs is being actively considered. Because these cells do not actively divide, retrovirus vectors cannot be used. Instead, delivery systems are being envisaged using vectors based on adenoviruses which are amphotrophic for such cells and do not require actively dividing cells.

Progress in understanding the molecular pathology of cancers has also suggested novel treatments. In the case of cancer due to inactivation of a tumor suppressor gene, transfer of the wild-type allele into tumor cells is theoretically sufficient to suppress growth of the cancer and has been demonstrated practically in certain cases [34]. Transfer of extra gene copies of the wild-type allele into somatic cells of individuals at risk is therefore expected to reduce the risk of cancer developing.

In the case of cancer due to oncogene activation, procedures for inactivating the expression of a specific oncogene have been considered. For example, antisense therapy envisages the inhibition of expression of an oncogene following transfer into the tumor cell of a suitably specific antisense RNA or oligodeoxynucleotide. By specifically binding to the oncogene mRNA the introduced antisense molecule can interfere with the translation of the oncogene mRNA and inhibit gene expression.

References

1. Ott, J. (1990) *Am. J. Hum. Genet.,* **46**, 219.
2. Bodmer, W.F., Bailey, C.J., Bodmer, J. *et al.* (1987) *Nature,* **328**, 614.
3. Kinzler, K.W., Nilbert, M.C., Vogelstein, B. *et al.* (1991) *Science,* **251**, 1366.
4. Nishisho, I., Nakamura, Y., Miyoshi, Y. *et al.* (1991) *Science,* **253**, 665.
5. Groden, J., Thliveris, A., Samowitz, W. *et al.* (1991) *Cell,* **66**, 589.
6. Levine, A.J., Momand, J. and Finlay, C.A. (1991) *Nature,* **351**, 453.
7. Fearon, E.R. and Vogelstein, B. (1990) *Cell,* **61**, 759.
8. Hobbs, H., Leitersdorf, E., Leffert, C.C., Cryer, D.R., Brown, M.S. and Goldstein, J.L. (1989) *J. Clin. Invest.,* **84**, 656.
9. Todd, J.A., Bell, J.I. and McDevitt, H.O. (1987) *Nature,* **329**, 599.
10. Miyazaki, T., Uno, M., Uehira, M. *et al.* (1990) *Nature,* **345**, 722.

11. Todd, J.A. (1990) *Immunol. Today,* **11**, 122.
12. Todd, J.A., Aitman, T.J., Cornall, R.J. *et al.* (1991) *Nature,* **351**, 542.
13. Cornall, R.J., Prins, J.-B., Todd, J.A. *et al.* (1991) *Nature,* **353**, 262.
14. St George-Hyslop, P.H., Haines, J.L., Farrer, L.A. *et al.* (1990) *Nature,* **347**, 194.
15. Pericak-Vance, M.A., Bebvout, J.L., Gaskell Jr, P.C. *et al.* (1991) *Am. J. Hum. Genet.,* **49**, 1034.
16. Goate, A., Chartier-Harlin, M.C., Mullan, M. *et al.* (1991) *Nature,* **349**, 704.
17. Zielenski, J., Rozmahel, R., Bozon, D. *et al.* (1991) *Genomics,* **10**, 214.
18. Anderson, M.P., Rich, D.P., Gregory, R.J., Smith, A.E. and Welsh, M.J. (1991) *Science,* **251**, 679.
19. Kartner, N., Hanrahan, J.W., Jensen, T.J. *et al.* (1991) *Cell,* **64**, 681.
20. Koenig, M., Monaco, A.P. and Kunkel, L.M. (1988). *Cell,* **53**, 219.
21. Love, D.R., Hill, D.F., Dickson, G. *et al.* (1989) *Nature,* **339**, 55.
22. Ervasti, J.M., Ohlendieck, K., Kahl, S.D., Gaver, M.G. and Campbell, K.P. (1990) *Nature,* **345**, 315.
23. Antonarakis, S.E. (1989) *New Engl. J. Med.,* **320**, 153.
24. Reiss, J. and Cooper, D.N. (1990) *Hum. Genet.,* **85**, 1.
25. Coutelle, C., Williams, C., Handyside, A. *et al.* (1989) *Br. Med. J.,* **299**, 22.
26. Abbs, S., Yau, S.C., Clark, S., Mathew, C.G. and Bobrow, M. (1991) *J. Med. Genet.,* **28**, 304.
27. Roberts, R.G., Bentley, D.R., Barby, T.F.M., Manners, E. and Bobrow, M. (1990) *Lancet,* **336**, 1523.
28. Hsiao, K.K., Scott, M., Foster, D. *et al.* (1990) *Science,* **250**, 1587.
29. Hammer, R.E., Maika, S.D., Richardson, J.A., Tang, J.-P. and Taurog, J.D. (1990) *Cell,* **63**, 1099.
30. Chisaka, O. and Capecchi, M.R. (1991) *Nature,* **350**, 473.
31. Friedmann, T. (1989) *Science,* **244**, 1275.
32. Shesely, E.G., Kim, H.-S.and Shehee, W.R. (1991) *Proc. Natl Acad. Sci. USA,* **88**, 4294.
33. Rosenberg, S.A., Aebersold, P., Cornetta, K. *et al.* (1990) *New Engl. J. Med.,* **323**, 570.
34. Baker, S.J., Markowitz, S., Fearon, E.R. *et al.* (1990) *Science,* **249**, 912.

Further reading

Weatherall, D.J. (1991) *The New Genetics and Clinical Practice,* 3rd edn. Oxford University Press, Oxford.

APPENDIX A. GLOSSARY

Allele: one of several alternative forms of a gene or DNA sequence at a particular chromosomal location (locus). At each locus an individual possesses two alleles, one inherited from the father and one from the mother.

Allele-specific oligonucleotide (ASO): a short oligonucleotide whose hybridization to a target sequence can be disrupted by a single base pair mismatch (see p.58).

Amplimer: oligonucleotide used as a primer of DNA synthesis in the polymerase chain reaction (see p. 63).

Annealing: association of complementary DNA (or RNA) strands to form a double helix.

Antisense strand: DNA strand of a gene which is used as a template for synthesis of mRNA.

Autosome: any chromosome other than the X and Y sex chromosomes.

Chromatid/chromosome: chromosomes as seen in dividing cells consist of two identical chromatids, joined at the centromere. See *Figure 1.3*, p. 6.

cDNA: complementary DNA, synthesized by the enzyme reverse transcriptase using mRNA as a template (see p. 52).

Coding DNA: DNA sequence which specifies the structure of a polypeptide or mature RNA (see p. 8).

Codon: a nucleotide triplet in mRNA which specifies one amino acid or a signal for terminating the synthesis of a polypeptide.

Contig: a series of overlapping DNA clones (see p. 85).

Cosmid: a vector for cloning DNA fragments in *E. coli* (see p. 67).

CpG island: short stretch of DNA rich in unmethylated CpG, often marking the position of a gene (see p. 87).

Cross-over: exchange of DNA between homologous chromosomes at meiosis (see p. 71).

Denaturation: dissociation of a double helix to give single-stranded DNA and/or RNA.

Diploid: cells containing two copies of the genome (46 chromosomes in human cells).

DNA library: collection of cell clones containing different recombinant DNA fragments (see p. 52).

Dominant: any trait which is expressed in a heterozygote.

Enhancer: DNA sequence element which stimulates transcription of a gene in a way not critically dependent on its position or orientation (see p. 14).

Exon: segment of a gene which is represented in the mature mRNA.

Gene cluster: multigene family whose members are clustered at a specific chromosomal location.

Gene conversion: a mechanism of non-reciprocal recombination, whereby the sequence of one gene comes to resemble that of another (see p. 40).

Genotype: (a) the genetic constitution of an individual; (b) the types of alleles found at a locus in an individual.

Germ-line: the gametes (egg and sperm cells) and precursor cells from which the gametes derive by cell division.

Haploid: cells containing one copy of the genome (23 chromosomes in human cells).

Haplotype: description of the types of alleles found at linked loci on a single chromosome (see p. 71).

Heterochromatin: region of a chromosome which remains condensed throughout the cell cycle and shows little or no gene expression (see p. 10).

Heteroduplex: double-stranded DNA in which the two strands are not perfectly complementary so that some bases are mis-paired.

Heterozygous: individuals are heterozygous at a locus if they have two different alleles at that locus.

Homologous chromosomes: chromosomes which carry the same loci but possibly different alleles, e.g. the two copies of chromosome 1 in a diploid cell. They are similar but not identical, one having been inherited from the father and the other from the mother.

Homozygous: individuals are homozygous at a locus if they exhibit two identical alleles at that locus.

Housekeeping gene: a gene coding for a basic cell function which is expressed in most cell types.

Hybridization assay: involves mixing single DNA (or RNA) strands from a labeled probe with those of a test DNA (or RNA) sample, then allowing complementary strands to anneal.

In-situ hybridization: hybridization of a labeled DNA or RNA fragment to a tissue section or chromosome spread on a microscope slide (see p. 81).

Intron: non-coding DNA which separates exons in a gene (see p. 16).

Linkage: tendency of alleles at two or more loci to be inherited together as a consequence of their physical proximity on a single chromosome.

Linkage disequilibrium: non-random association of alleles at linked loci.

Locus: unique chromosomal location defining the position of an individual gene or DNA sequence.

Lod score: a measure of the likelihood of genetic linkage between loci (see p. 77).

Marker: a polymorphic DNA or protein sequence whose transmission can be followed through a pedigree.

Meiosis: reductive cell division occurring exclusively in testes and ovaries and resulting in the production of haploid sperm and egg cells.

Microsatellite: a stretch of tandemly repeated very short DNA sequences (see p. 25).

Minisatellite: a stretch of tandemly repeated short DNA sequences (see p. 24).

Mitosis: normal cell division, as distinct from meiosis.

Mosaic: a genetic mosaic is an individual who has two or more genetically different cell lines derived from a single zygote (see p. 104).

Mutation: a heritable alteration in the DNA sequence.

Polymerase chain reaction (PCR): an *in-vitro* method of DNA cloning (see p. 62).

Polymorphism: existence of two or more alleles at significant frequencies in the population.

Probe: a DNA or RNA fragment which has been labeled in some way and used in a hybridization assay to identify related DNA or RNA sequences.

Promoter: the DNA sequence elements upstream of a gene, to which RNA polymerase binds in order to initiate transcription (see p. 14).

Pseudogene: a DNA sequence clearly related to a non-allelic functional gene, but which is itself non-functional.

Rare cutter: restriction nuclease which cuts DNA infrequently because the sequence it recognizes is large and/or contains one or more CpGs (see p. 68).

Repetitive DNA: DNA sequence present in many identical or very similar copies in the genome (see pp. 18–26).

Restriction site: short DNA sequence often 4–8 bp long which is recognized by a restriction nuclease.

Restriction site polymorphism (RSP): polymorphism involving two alleles which differ in possessing or lacking a specific restriction site.

Satellite DNA: highly repetitive non-transcribed DNA (see p. 23).

Sense strand: DNA strand of a gene which is complementary to the antisense strand. Its base sequence is identical to the transcribed RNA sequence, except that the base thymine in DNA corresponds to uracil in RNA.

Silencer: combination of short DNA sequence elements which suppress the transcription of a gene.

Somatic cell: any cell in the body except the gametes.

Somatic cell hybrid: artificially constructed cell in which chromosomes of one species (usually human) have been stably introduced into cells of another species (usually rodent) (see p. 82).

Splicing: RNA splicing is removal of introns during mRNA processing (see p. 16). DNA splicing is a much rarer event (see p. 43).

Syntenic: on the same chromosome.

Transgenic animal: an animal in which artificially introduced foreign DNA becomes stably incorporated into the germ-line.

Translocation: transfer of chromosomal regions between non-homologous chromosomes.

Untranslated sequences: non-coding sequences at the 5' and 3' ends of mRNA.

Unequal cross-over: recombination between improperly paired sequences on two homologous chromosomes.

Unequal sister chromatid exchange: recombination between improperly paired sequences on the sister chromatids of a single chromosome.

Yeast artificial chromosome (YAC): a vector used for cloning large foreign DNA fragments in yeast cells (see p. 67).

Zoo blot: a Southern blot containing DNA samples from different species (see p. 86).

INDEX